LATENT ROOTS
AND
LATENT VECTORS

S. J. HAMMARLING

B.Sc.

Lecturer in Numerical Analysis, Enfield College of Technology

Latent Roots and Latent Vectors

THE UNIVERSITY OF TORONTO PRESS

First published in Canada and the United States by
The University of Toronto Press

ISBN 0 8020 1709 6

Published in Great Britain by
Adam Hilger Ltd

Printed in Great Britain by John Wright & Sons Ltd, Bristol

To

V. M. and P. H.

PREFACE

This book is concerned with the latent roots† and latent vectors of matrices whose elements are real. Its main purpose is to discuss some of the methods available for finding latent roots and vectors. The methods presented include not only those that are the most useful in practice, but some chosen because of the interesting ideas they present and the help they give to a general understanding of the latent root and vector problem. I have attempted throughout to introduce the material in as uncomplicated a manner as possible.

One of the reasons for the book is the importance of latent roots in a wide field of applications, for example, nuclear and atomic physics, statistical analysis, aeronautics, structural analysis, design of control systems, vibration theory, and the convergence and stability of various numerical methods. I hope this book may be of use to people working in such areas.

I have assumed a knowledge of elementary matrix and determinant theory. The first chapter gives many of the theorems required in the later chapters. Schur's theorem, that every matrix is similar to a triangular matrix, is given early on in the belief that this has often not been used to its full potential. It is also of great practical importance since the triangular form can be obtained by stable methods such as the **Q–R** algorithm, in contrast to forms such as the Jordan canonical form which can only be obtained by unstable methods. The second chapter presents just four of the many uses of latent roots and vectors and these reflect only my own interest in the subject. The remainder of the book is devoted to methods of finding latent roots and vectors.

I have attempted to illustrate all the methods with simple examples that can be easily followed and understood. In many cases, for the purpose of illustration, the examples and exercises are so constructed that exact arithmetic is possible. Obviously this is not the case in practical examples, so it must be borne in mind that these particular examples may not reflect the numerical problems that can arise. In practice a knowledge of the condition of a problem with respect to its solution is desirable. By condition I mean a measure of the sensitivity of a solution with respect to changes in the original data. This measure is clearly important since we are unlikely to have exact data and also we shall have to introduce rounding errors in computing a solution. A method which leads to an ill-conditioned problem is unstable, and of the methods discussed those of Danilevsky, Krylov and Lanczos can be said to be unstable since they are either directly or indirectly connected with the Frobenius form which can be extremely ill-conditioned with respect to its latent roots. For a full discussion of this type of problem and for a detailed error analysis of most of the methods the reader can but be referred to J. H. Wilkinson's *The Algebraic Eigenvalue Problem.*

† Often referred to as eigenvalues, characteristic values, characteristic roots and proper values.

My book grew out of a project undertaken whilst I was a student on the Mathematics for Business course at Enfield College of Technology, and my grateful thanks are due to the mathematics staff concerned with that course. I must also thank the staff of Adam Hilger Ltd, Mr. S. Millward, for much help and encouragement, and my wife Pam, who bravely typed the original manuscript and filled in all those weird symbols.

SVEN HAMMARLING

Enfield

March, 1970

CONTENTS

NOTATION

Upper case letters have been used almost exclusively for matrices and vectors, and lower case and Greek letters for scalars. λ has been used only to represent a latent root. The matrix $\mathbf{A}$ is generally the matrix of prime interest, and, unless otherwise specified, is an $n \times n$ matrix containing only real elements.

$\|\mathbf{A}\|$	Determinant of the matrix $\mathbf{A}$.
$\mathbf{A}^{-1}$	Inverse of the matrix $\mathbf{A}$.
$\mathbf{A}^T$	Transpose of the matrix $\mathbf{A}$.
$\mathbf{A}^*$	Matrix whose elements are the complex conjugate of $\mathbf{A}^T$.
$\|\alpha\|$	Modulus of α. No confusion should arise with $\|\mathbf{A}\|$.
$\mathbf{I}$	The unit matrix.

Chapter 1

LATENT ROOTS AND LATENT VECTORS

1.1 Latent Roots and Latent Vectors

A *latent root* of a square matrix **A** is a number, λ, that satisfies the equation,

$$\mathbf{AX} = \lambda\mathbf{X} \tag{1.1}$$

where **X** is a column vector and is known as a *latent vector* of **A**.

The values of λ that satisfy equation (1.1) when $\mathbf{X} \neq 0$ are given by solving the determinantal equation, called the *characteristic equation* of **A**,

$$|\mathbf{A} - \lambda\mathbf{I}| = 0 \tag{1.2}$$

since (1.1) may be written as

$$(\mathbf{A} - \lambda\mathbf{I})\,\mathbf{X} = 0$$

and since, if $|\mathbf{A} - \lambda\mathbf{I}| \neq 0$, then $(\mathbf{A} - \lambda\mathbf{I})^{-1}$ exists, it follows that

$$(\mathbf{A} - \lambda\mathbf{I})^{-1}(\mathbf{A} - \lambda\mathbf{I})\,\mathbf{X} = \mathbf{IX} = \mathbf{X} = 0$$

The solution of equation (1.2) as it stands involves the evaluation of an $n \times n$ determinant and the extraction of n roots from the resulting polynomial in λ. If the latent vectors are also required, we shall have to solve the n equations of (1.1) for each value of λ. Since the determinant of (1.2) is not wholly arithmetic, its evaluation will involve of the order of $n!$ calculations. The general solution of the problem in this form is clearly impracticable.

Example 1.1

$$\mathbf{A} = \begin{pmatrix} 3 & 1 \\ 2 & 4 \end{pmatrix}$$

Hence

$$|\mathbf{A} - \lambda\mathbf{I}| = \begin{vmatrix} 3-\lambda & 1 \\ 2 & 4-\lambda \end{vmatrix} = \lambda^2 - 7\lambda + 10 = (\lambda - 5)(\lambda - 2) = 0$$

So the latent roots of **A** are $\lambda_1 = 5$ and $\lambda_2 = 2$. From $\mathbf{AX} = \lambda\mathbf{X}$ we get

$$3x + y = \lambda x$$

$$2x + 4y = \lambda y$$

When $\lambda = 5$,

$$3x + y = 5x$$

$$2x + 4y = 5y$$

Hence

$$y = 2x$$

When $\lambda = 2$,

$$3x + y = 2x$$

$$2x + 4y = 2y$$

Hence

$$y = -x$$

The latent vectors of **A** are any vectors of the form

$$\mathbf{X}_1 = k\begin{pmatrix} 1 \\ 2 \end{pmatrix} \quad \text{and} \quad \mathbf{X}_2 = k\begin{pmatrix} 1 \\ -1 \end{pmatrix}$$

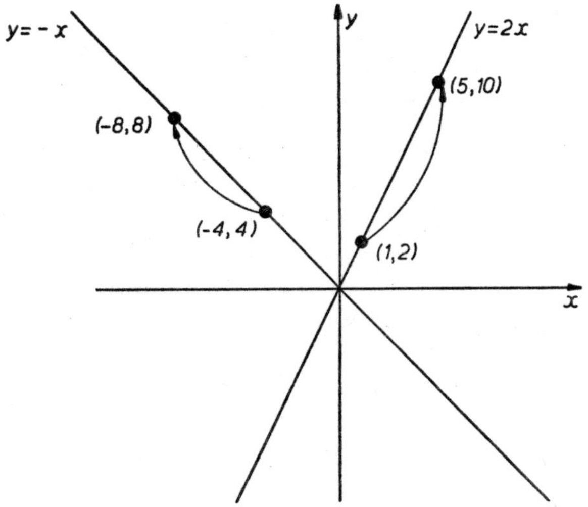

Fig. 1

Geometrically, we have found those vectors which remain unaltered in direction when they are transformed by the matrix **A**. The latent root measures the change in magnitude of the latent vector. See Fig. 1.

Various results and theorems that will be needed in later work are now given.

1.2 Similar Matrices

Similar matrices and similarity transformations will play an important role in much of the work in this book.

Two matrices $\mathbf{A}$ and $\mathbf{B}$ are said to be *similar* if there exists a matrix $\mathbf{C}$ such that

$$\mathbf{B} = \mathbf{C}^{-1}\mathbf{A}\mathbf{C}$$

The transformation from $\mathbf{A}$ to $\mathbf{B}$ is called a *similarity transformation.* If $\mathbf{C}$ is a matrix such that

$$\mathbf{C}^T = \mathbf{C}^{-1}$$

it is called an *orthogonal* matrix, and the similarity transformation is said to be an *orthogonal transformation.* An important orthogonal matrix is given by

$$\mathbf{R} = \begin{pmatrix} \cos\theta & -\sin\theta \\ \sin\theta & \cos\theta \end{pmatrix}$$

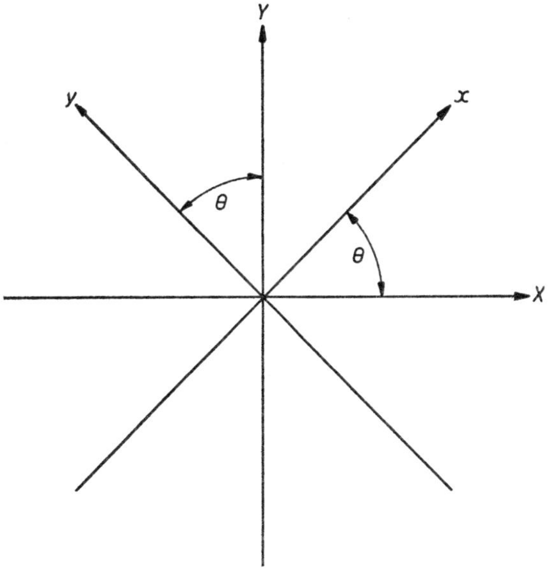

Fig. 2

which has the effect of rotating the x and y axes through an angle $-\theta$ into new axes X and Y for

$$\begin{pmatrix} \cos\theta & -\sin\theta \\ \sin\theta & \cos\theta \end{pmatrix} \begin{pmatrix} x \\ y \end{pmatrix} = \begin{pmatrix} X \\ Y \end{pmatrix}$$

gives

$$X = \cos\theta . x - \sin\theta . y$$
$$Y = \sin\theta . x + \cos\theta . y$$

which can be shown by elementary algebra to give the required transformation.

Theorem 1.1

Similar matrices have the same characteristic equation.

Proof

$$\mathbf{B} = \mathbf{C}^{-1}\mathbf{AC}$$

Hence

$$|\mathbf{B}-\lambda\mathbf{I}| = |\mathbf{C}^{-1}\mathbf{AC}-\lambda\mathbf{I}| = |\mathbf{C}^{-1}\mathbf{AC}-\lambda\mathbf{C}^{-1}\mathbf{IC}|$$
$$= |\mathbf{C}^{-1}(\mathbf{A}-\lambda\mathbf{I})\mathbf{C}| = |\mathbf{C}^{-1}||\mathbf{A}-\lambda\mathbf{I}||\mathbf{C}|$$
$$= |\mathbf{C}^{-1}||\mathbf{C}||\mathbf{A}-\lambda\mathbf{I}|$$
$$= |\mathbf{A}-\lambda\mathbf{I}|$$

Theorem 1.2

If $\mathbf{B} = \mathbf{C}^{-1}\mathbf{AC}$ and $\mathbf{X}$ and $\mathbf{Y}$ are the respective latent vectors of $\mathbf{A}$ and $\mathbf{B}$ corresponding to the latent root λ, then,

$$\mathbf{CY} = \mathbf{X}$$

Proof

$$\mathbf{BY} = \lambda\mathbf{Y}$$

Hence

$$\mathbf{CBY} = \lambda\mathbf{CY}$$

But

$$\mathbf{CB} = \mathbf{AC}$$

so that

$$\mathbf{ACY} = \lambda\mathbf{CY}$$

which gives

$$\mathbf{CY} = \mathbf{X}$$

Theorem 1.3

Every matrix is similar to a triangular matrix, i.e. a matrix having zero in each position either above or below its leading diagonal.

Before proving this important theorem we need some intermediate results.

We denote by $\mathbf{C}^*$ the matrix whose elements are the complex conjugate of $\mathbf{C}^T$. If $\mathbf{C}$ is such that

$$\mathbf{C}^* = \mathbf{C}^{-1}$$

it is called a *unitary matrix*. If the elements of $\mathbf{C}$ are all real, then, of course, it is orthogonal.

Theorem 1.4

The elements on the leading diagonal of a triangular matrix are its latent roots.

Proof

Let

$$\mathbf{A} = \begin{pmatrix} a_{11} & 0 & \dots & 0 \\ a_{21} & a_{22} & \dots & 0 \\ \vdots & \vdots & \ddots & \vdots \\ a_{n1} & a_{n2} & \dots & a_{nn} \end{pmatrix}$$

Then

$$|\mathbf{A}-\lambda\mathbf{I}| = (a_{11}-\lambda)(a_{22}-\lambda)\dots(a_{nn}-\lambda) = 0$$

The case of an upper triangular matrix is equally simple.

We now restate theorem 1.3 more strongly and prove it by induction.

Theorem 1.5 (*Schur's theorem*)

For every matrix $\mathbf{A}$, there exists a unitary matrix $\mathbf{C}$ such that

$$\mathbf{B} = \mathbf{C}^{-1}\mathbf{A}\mathbf{C}$$

where $\mathbf{B}$ is triangular.

Proof

When $\mathbf{A}_1 = (a_{11})$ the theorem is clearly true.

Suppose that the theorem is true when $\mathbf{A}_n$ is an $n \times n$ matrix.

Let the latent roots of $\mathbf{A}_{n+1}$ be $\lambda_1, \lambda_2, \dots, \lambda_{n+1}$, and $\mathbf{X}_i$ be the latent vector corresponding to λ_i normalized so that

$$\mathbf{X}_i^* \mathbf{X}_i = 1$$

Further, let us choose a matrix $\mathbf{C}_1$ having $\mathbf{X}_1$ as its first column and with its remaining columns such that $\mathbf{C}_1$ is unitary. Then we find that, since $\mathbf{C}_1$ is unitary,

$$\mathbf{C}_1^* \mathbf{A}_{n+1} \mathbf{C}_1 = \mathbf{C}_1^{-1} \mathbf{A}_{n+1} \mathbf{C}_1 = \left(\begin{array}{c|cccc} \lambda_1 & c_1 & c_2 & \dots & c_n \\ \hline 0 & & & & \\ 0 & & & & \\ \vdots & & & \mathbf{A}_n & \\ 0 & & & & \end{array}\right) = \mathbf{B}_1$$

where $\mathbf{A}_n$ is an $n \times n$ matrix. The characteristic equation of $\mathbf{B}_1$ is given by

$$|\mathbf{B}_1 - \lambda\mathbf{I}| = (\lambda_1 - \lambda)|\mathbf{A}_n - \lambda\mathbf{I}| = 0$$

and hence the latent roots of $\mathbf{A}_n$ are $\lambda_2, \lambda_3, \ldots, \lambda_{n+1}$. Our inductive hypothesis asserts that we can find a unitary matrix $\mathbf{C}_n$ such that

$$\mathbf{C}_n^{-1}\mathbf{A}_n\mathbf{C}_n = \begin{pmatrix} \lambda_2 & d_{12} & d_{13} & \ldots & d_{1n} \\ 0 & \lambda_3 & d_{23} & \ldots & d_{2n} \\ 0 & 0 & \lambda_4 & \ldots & d_{3n} \\ \vdots & \vdots & \vdots & & \vdots \\ 0 & 0 & 0 & \ldots & \lambda_{n+1} \end{pmatrix}$$

Let $\mathbf{C}_{n+1}$ be the unitary matrix given by

$$\mathbf{C}_{n+1} = \left(\begin{array}{c|ccc} 1 & 0 \; 0 & \ldots & 0 \\ \hline 0 & & & \\ \vdots & & & \\ 0 & & \mathbf{C}_n & \\ 0 & & & \end{array}\right)$$

We now find that

$$(\mathbf{C}_1\mathbf{C}_{n+1})^{-1}\mathbf{A}_{n+1}(\mathbf{C}_1\mathbf{C}_{n+1}) = \mathbf{C}_{n+1}^{-1}(\mathbf{C}_1^{-1}\mathbf{A}_{n+1}\mathbf{C}_1)\mathbf{C}_{n+1} = \mathbf{C}_{n+1}^{-1}\mathbf{B}_1\mathbf{C}_{n+1}$$

$$= \begin{pmatrix} \lambda_1 & c_{12} & c_{13} & \ldots & c_{1,n+1} \\ 0 & \lambda_2 & c_{23} & \ldots & c_{2,n+1} \\ 0 & 0 & \lambda_3 & \ldots & c_{3,n+1} \\ \vdots & \vdots & \vdots & & \vdots \\ 0 & 0 & 0 & \ldots & \lambda_{n+1} \end{pmatrix} = \mathbf{B}$$

Putting $\mathbf{C} = \mathbf{C}_1\mathbf{C}_{n+1}$, the theorem is proved.†

1.3 Theorems Concerning Latent Roots

Theorem 1.6

If λ and $\mathbf{X}$ are a corresponding latent root and vector of $\mathbf{A}$, then λ^m and $\mathbf{X}$ are a corresponding latent root and vector of $\mathbf{A}^m$, m being an integer. (The case of negative m is valid, of course, only if $\mathbf{A}^{-1}$ exists.)

Proof

We have

$$\mathbf{AX} = \lambda\mathbf{X}$$

† Working through exercise 2.13(i) may help to follow the proof.

and, if $\mathbf{A}^{-1}$ exists,

$$\mathbf{X} = \lambda \mathbf{A}^{-1}\mathbf{X}$$

Assume that the theorem holds for $m = r$ so that

$$\mathbf{A}^r\mathbf{X} = \lambda^r\mathbf{X}$$

which gives

$$\mathbf{A}^{r+1}\mathbf{X} = \lambda^r\mathbf{AX} = \lambda^{r+1}\mathbf{X}$$

Also, when $\mathbf{A}^{-1}$ exists,

$$\mathbf{A}^{r-1}\mathbf{X} = \lambda^r\mathbf{A}^{-1}\mathbf{X} = \lambda^{r-1}\mathbf{X}$$

Hence by induction the theorem is proved.

Theorem 1.7

$\mathbf{A}^m \to 0$ as $m \to \infty$ if and only if $|\lambda| < 1$ for all λ.

Proof

First assume that $\mathbf{A}^m \to 0$ as $m \to \infty$.

From theorem 1.6 we have

$$\mathbf{A}^m\mathbf{X} = \lambda^m\mathbf{X}$$

so that, if $\mathbf{A}^m \to 0$ as $m \to \infty$, $\mathbf{A}^m\mathbf{X} \to 0$. Hence $\lambda^m\mathbf{X} \to 0$, which means that $|\lambda| < 1$.

Now suppose that $|\lambda| < 1$ for all λ.

We saw in theorem 1.5 that we can put

$$\mathbf{A} = \mathbf{CB}_1\mathbf{C}^{-1}$$

where $\mathbf{B}_1$ is triangular. It is easy to show by induction that we have

$$\mathbf{A}^m = \mathbf{CB}_1^m\mathbf{C}^{-1}$$

Now, $\mathbf{B}_1^m$ is triangular and has $\lambda_1^m, \lambda_2^m, \ldots, \lambda_n^m$ as the elements on its leading diagonal, so that in the limit as $m \to \infty$, we get

$$\mathbf{B}_1^m \to \mathbf{B} = \begin{pmatrix} 0 & b_{12} & b_{13} & \cdots & b_{1n} \\ 0 & 0 & b_{23} & \cdots & b_{2n} \\ 0 & 0 & 0 & \cdots & b_{3n} \\ \vdots & \vdots & \vdots & & \vdots \\ 0 & 0 & 0 & \cdots & 0 \end{pmatrix}$$

but it is clear that $\mathbf{B}^n = 0 = \lim_{m\to\infty} \mathbf{B}_1^{mn} = \lim_{m\to\infty} \mathbf{B}_1^m$. Hence we must have $\mathbf{B} = 0$ and $\mathbf{B}_1^m \to 0$ as $m \to \infty$.

Since

$$\mathbf{A}^m = \mathbf{CB}_1^m\mathbf{C}$$

$$\mathbf{A}^m \to 0 \quad \text{as } \mathbf{m} \to \infty$$

which is the required result.

Theorem 1.8 (*Cayley–Hamilton theorem*)

A matrix satisfies its own characteristic equation, i.e. if the characteristic equation of **A** is

$$\lambda^n - p_1\lambda^{n-1} - p_2\lambda^{n-2} - \dots - p_{n-1}\lambda - p_n = 0$$

then

$$\mathbf{A}^n - p_1\mathbf{A}^{n-1} - p_2\mathbf{A}^{n-2} - \dots - p_{n-1}\mathbf{A} - p_n\mathbf{I} = 0$$

Proof

Consider the matrix **B**, which is the adjoint of the matrix $(\mathbf{A}-\lambda\mathbf{I})$. That is,

$$\mathbf{B}(\mathbf{A}-\lambda\mathbf{I}) = |\mathbf{A}-\lambda\mathbf{I}|.\mathbf{I} \qquad (1.3)$$

Each element of **B** is a co-factor of the determinant $|\mathbf{A}-\lambda\mathbf{I}|$, and hence is a polynomial in λ of degree not greater than $n-1$. This means that we can find matrices $\mathbf{B}_1, \mathbf{B}_2, \dots, \mathbf{B}_n$ independent of λ such that

$$\mathbf{B} = \mathbf{B}_1\lambda^{n-1} + \mathbf{B}_2\lambda^{n-2} + \dots + \mathbf{B}_{n-1}\lambda + \mathbf{B}_n$$

Hence (1.3) becomes

$$(\mathbf{B}_1\lambda^{n-1} + \mathbf{B}_2\lambda^{n-2} + \dots + \mathbf{B}_{n-1}\lambda + \mathbf{B}_n)(\mathbf{A}-\lambda\mathbf{I}) = (\lambda^n - p_1\lambda^{n-1} - \dots - p_{n-1}\lambda - p_n).\mathbf{I}$$

and equating coefficients of λ we get

$$\begin{aligned}
\mathbf{B}_n\mathbf{A} \qquad &= -\,p_n\mathbf{I}\\
\mathbf{B}_{n-1}\mathbf{A} - \mathbf{B}_n &= -p_{n-1}\mathbf{I}\\
\vdots \qquad \vdots \quad &\qquad \vdots\\
\mathbf{B}_1\mathbf{A} \ - \mathbf{B}_2 &= -\,p_1\mathbf{I}\\
-\mathbf{B}_1 &= \mathbf{I}
\end{aligned}$$

Post-multiplying the first of these by **I**, the second by **A**, the third by $\mathbf{A}^2, \dots,$ the nth by $\mathbf{A}^{n-1}$, the $(n+1)$th by $\mathbf{A}^n$ and adding we get

$$0 = -p_n\mathbf{I} - p_{n-1}\mathbf{A} - \dots - p_1\mathbf{A}^{n-1} + \mathbf{A}^n$$

or

$$\mathbf{A}^n - p_1\mathbf{A}^{n-1} - p_2\mathbf{A}^{n-2} - \dots - p_{n-1}\mathbf{A} - p_n\mathbf{I} = 0$$

as required.

Theorem 1.9

The latent roots of the transpose of **A** are the same as those of **A**.

Proof

The latent roots of **A** are given by

$$|\mathbf{A}-\lambda\mathbf{I}| = 0$$

Since $|\mathbf{A}-\lambda\mathbf{I}|$ is formed by subtracting only from the leading diagonal of $\mathbf{A}$ we have

$$|\mathbf{A}^T-\lambda\mathbf{I}| = |\mathbf{A}-\lambda\mathbf{I}|^T = |\mathbf{A}-\lambda\mathbf{I}| = 0$$

and the theorem is proved.

Theorem 1.10 (*Gerschgorin's theorem*)

The modulus of any latent root of a matrix does not exceed the largest sum of the moduli of the elements of any one row or column.

Proof

Let the components of the latent vector $\mathbf{X}$ of $\mathbf{A}$ be $x_1, x_2, \ldots, x_n$.

Then, since

$$\mathbf{AX} = \lambda\mathbf{X}$$

we have

$$\begin{aligned}
a_{11}x_1 + a_{12}x_2 + \ldots + a_{1n}x_n &= \lambda x_1 \\
a_{21}x_1 + a_{22}x_2 + \ldots + a_{2n}x_n &= \lambda x_2 \\
\vdots \qquad \vdots \qquad\quad \vdots \quad &\quad \vdots \\
a_{n1}x_1 + a_{n2}x_2 + \ldots + a_{nn}x_n &= \lambda x_n
\end{aligned}$$

Let $|x_r| \geqslant |x_i|$ for all i, then selecting the rth equation above we get

$$\lambda = a_{r1}.\frac{x_1}{x_r} + a_{r2}.\frac{x_2}{x_r} + \ldots + a_{rn}.\frac{x_n}{x_r}$$

so that

$$|\lambda| \leqslant |a_{r1}| + |a_{r2}| + \ldots + |a_{rn}|$$

because $|x_i/x_r| \leqslant 1$ for all i.

The column case follows from theorem 1.9.

Theorem 1.11

The sum of the latent roots of $\mathbf{A}$ is equal to its trace.

Proof

The latent roots of $\mathbf{A}$ are given by

$$|\mathbf{A}-\lambda\mathbf{I}| = 0$$

or

$$\begin{vmatrix} a_{11}-\lambda & a_{12} & \ldots & a_{1n} \\ a_{21} & a_{22}-\lambda & \ldots & a_{2n} \\ \vdots & \vdots & & \vdots \\ a_{n1} & a_{n2} & \ldots & a_{nn}-\lambda \end{vmatrix} = 0$$

which gives, say,

$$\lambda^n - p_1\lambda^{n-1} - \ldots - p_{n-1}\lambda - p_n = 0$$

and if the theorem is true we must have

$$p_1 = -(a_{11} + a_{22} + \ldots + a_{nn})$$

We see that, if $\mathbf{A}_1 = [a_{11}]$, the theorem is true. Assume that the theorem is true for an $(n-1)\times(n-1)$ matrix.

Then, if we expand the above determinant by its last row, $(a_{nn} - \lambda)$ is the only element to contribute to the coefficient of λ^{n-1}, so that if our inductive hypothesis holds, this element times its co-factor yields

$$(a_{nn} - \lambda)\{\lambda^{n-1} - (a_{11} + a_{22} + \ldots + a_{n-1,n-1})\lambda^{n-2} - \ldots - p'_n\}$$

which gives as the coefficient of λ^{n-1}

$$p_1 = -(a_{11} + a_{22} + \ldots + a_{nn})$$

(The above gives $-p_1$ but also $-\lambda^n$, and since we equate the characteristic equation to zero we are justified in changing the sign.) So by induction the theorem is proved.

Theorem 1.12

The product of the latent roots of $\mathbf{A}$ is equal to $|\mathbf{A}|$.

Proof

From theorem 1.5 we can express $\mathbf{A}$ as

$$\mathbf{A} = \mathbf{CBC}^{-1}$$

where $\mathbf{B}$ is triangular. Hence

$$|\mathbf{A}| = |\mathbf{C}|\,|\mathbf{B}|\,|\mathbf{C}^{-1}| = |\mathbf{C}|\,|\mathbf{C}^{-1}|\,|\mathbf{B}| = |\mathbf{B}| = \lambda_1\lambda_2\ldots\lambda_n$$

as required.

Theorem 1.13

If λ is a latent root of $\mathbf{A}$, and $f(\mathbf{A})$ is a polynomial in $\mathbf{A}$, then $f(\lambda)$ is a latent root of $f(\mathbf{A})$.

Proof

Let $f(\mathbf{A}) = a_n\mathbf{A}^n + a_{n-1}\mathbf{A}^{n-1} + \ldots + a_1\mathbf{A} + a_0\mathbf{I}$.

Then

$$f(\mathbf{A})\mathbf{X} = a_n\mathbf{A}^n\mathbf{X} + a_{n-1}\mathbf{A}^{n-1}\mathbf{X} + \ldots + a_1\mathbf{AX} + a_0\mathbf{X}$$

so that if $\mathbf{X}$ is a latent vector of $\mathbf{A}$, from theorem 1.3 we get

$$\begin{aligned} f(\mathbf{A})\mathbf{X} &= a_n\lambda^n\mathbf{X} + a_{n-1}\lambda^{n-1}\mathbf{X} + \ldots + a_1\lambda\mathbf{X} + a_0\mathbf{X} \\ &= (a_n\lambda^n + a_{n-1}\lambda^{n-1} + \ldots + a_1\lambda + a_0)\mathbf{X} \\ &= f(\lambda)\mathbf{X} \end{aligned}$$

as required.

Theorem 1.14

If $\mathbf{A} = \left(\begin{array}{c|c} \mathbf{D}_1 & \mathbf{D}_2 \\ \hline 0 & \mathbf{D}_3 \end{array}\right)$, $\mathbf{D}_1$ and $\mathbf{D}_3$ being square,

then,

$$|\mathbf{A}-\lambda\mathbf{I}| = |\mathbf{D}_1-\lambda\mathbf{I}|\,|\mathbf{D}_3-\lambda\mathbf{I}|$$

Proof

Clearly it suffices to show that $|\mathbf{A}| = |\mathbf{D}_1|\,|\mathbf{D}_3|$. Firstly we see that if $\mathbf{D}_1$ contains just one element, say a, then expanding $|\mathbf{A}|$ by the first column we get

$$|\mathbf{A}| = a|\mathbf{D}_3| = |\mathbf{D}_1|\,|\mathbf{D}_3|$$

Assume that the theorem is true when $\mathbf{D}_1$ is an $(n-1)\times(n-1)$ matrix. Now let

$$|\mathbf{A}| = \begin{vmatrix} a_{11} & a_{12} & \dots & a_{1n} & c_{11} & c_{12} & \dots & c_{1m} \\ a_{21} & a_{22} & \dots & a_{2n} & c_{21} & c_{22} & \dots & c_{2m} \\ \vdots & \vdots & & \vdots & \vdots & \vdots & & \vdots \\ a_{n1} & a_{n2} & \dots & a_{nn} & c_{n1} & c_{n2} & \dots & c_{nm} \\ 0 & 0 & \dots & 0 & b_{11} & b_{12} & \dots & b_{1m} \\ 0 & 0 & \dots & 0 & b_{21} & b_{22} & \dots & b_{2m} \\ \vdots & \vdots & & \vdots & \vdots & \vdots & & \vdots \\ 0 & 0 & \dots & 0 & b_{m1} & b_{m2} & \dots & b_{mm} \end{vmatrix} = \left|\begin{array}{c|c} \mathbf{D}_1 & \mathbf{D}_2 \\ \hline 0 & \mathbf{D}_3 \end{array}\right|$$

Expanding $|\mathbf{A}|$ by the first column we get

$$|\mathbf{A}| = a_{11}\left|\begin{array}{c|c} \mathbf{M}_{11} & \mathbf{X}_1 \\ \hline 0 & \mathbf{D}_3 \end{array}\right| - a_{21}\left|\begin{array}{c|c} \mathbf{M}_{21} & \mathbf{X}_2 \\ \hline 0 & \mathbf{D}_3 \end{array}\right| + \dots + (-1)^{n+1}a_{n1}\left|\begin{array}{c|c} \mathbf{M}_{n1} & \mathbf{X}_n \\ \hline 0 & \mathbf{D}_3 \end{array}\right|$$

where $\mathbf{M}_{ij}$ is the minor of a_{ij} with respect to $\mathbf{D}_1$. But $\mathbf{M}_{ij}$ is an $(n-1)\times(n-1)$ group of elements, hence by our inductive assumption

$$\begin{aligned} |\mathbf{A}| &= a_{11}\mathbf{M}_{11}|\mathbf{D}_3| - a_{21}\mathbf{M}_{21}|\mathbf{D}_3| + \dots + (-1)^{n+1}a_{n1}\mathbf{M}_{n1}|\mathbf{D}_3| \\ &= [a_{11}\mathbf{M}_{11} - a_{21}\mathbf{M}_{21} + \dots + (-1)^{n+1}a_{n1}\mathbf{M}_{n1}]\,|\mathbf{D}_3| = |\mathbf{D}_1|\,|\mathbf{D}_3| \end{aligned}$$

and so by induction the theorem is proved.

1.4 Theorems Concerning Latent Vectors

Theorem 1.15

If $\mathbf{A}$ has r distinct latent roots, then it has at least r linearly independent latent vectors.

Proof

From theorem 1.5 we can put $\mathbf{A} = \mathbf{CBC}^{-1}$, where $\mathbf{B}$ is triangular. From the construction of that theorem it is clear that the latent roots of $\mathbf{A}$ can be made to appear in any order on the leading diagonal of $\mathbf{B}$. Suppose that the r distinct latent roots $\lambda_1, \lambda_2, \ldots, \lambda_r$ appear in the first r positions. Then, if $\mathbf{Y}$ is a latent vector of $\mathbf{B}$ having elements $y_1, y_2, \ldots, y_n$, from $\mathbf{BY} = \lambda\mathbf{Y}$ we get

$$\begin{aligned}
\lambda_1 y_1 + b_{12} y_2 + b_{13} y_3 + \ldots + b_{1n} y_n &= \lambda y_1 \\
\lambda_2 y_2 + b_{23} y_3 + \ldots + b_{2n} y_n &= \lambda y_2 \\
\lambda_3 y_3 + \ldots + b_{3n} y_n &= \lambda y_3 \\
\vdots \qquad & \quad \vdots \\
\lambda_n y_n &= \lambda y_n
\end{aligned}$$

Suppose that $\lambda = \lambda_p$, where $p \leqslant r$, and we put $y_{p+1} = y_{p+2} = \ldots = y_p = 0$. We then get the equations

$$\begin{aligned}
(\lambda_1 - \lambda_p) y_1 + b_{12} y_2 + b_{13} y_3 + \ldots + b_{1p} y_p &= 0 \\
(\lambda_2 - \lambda_p) y_2 + b_{23} y_3 + \ldots + b_{2p} y_p &= 0 \\
(\lambda_3 - \lambda_p) y_3 + \ldots + b_{3p} y_p &= 0 \\
\vdots \qquad & \quad \vdots \\
(\lambda_p - \lambda_p) y_p &= 0
\end{aligned}$$

Since the last equation is $0 = 0$, we are left with $p-1$ equations in p unknowns. If we fix $y_p \neq 0$, we then have $p-1$ equations in $p-1$ unknowns and since none of the values $\lambda_1, \lambda_2, \ldots, \lambda_{p-1}$ are equal to λ_p, the rank of the equations is $p-1$ so that we can solve uniquely for $y_1, y_2, \ldots, y_{p-1}$. We have now clearly found a particular solution to the original set of equations, so that the r latent vectors of $\mathbf{B}$ corresponding to $\lambda_1, \lambda_2, \ldots, \lambda_r$ are respectively

$$\begin{pmatrix} y_1 \\ 0 \\ 0 \\ \vdots \\ 0 \\ 0 \\ 0 \\ \vdots \\ 0 \end{pmatrix}, \begin{pmatrix} z_{21} \\ y_2 \\ 0 \\ \vdots \\ 0 \\ 0 \\ 0 \\ \vdots \\ 0 \end{pmatrix}, \begin{pmatrix} z_{31} \\ z_{32} \\ y_3 \\ \vdots \\ 0 \\ 0 \\ 0 \\ \vdots \\ 0 \end{pmatrix}, \ldots, \begin{pmatrix} z_{r1} \\ z_{r2} \\ z_{r3} \\ \vdots \\ z_{r,r-1} \\ y_r \\ 0 \\ \vdots \\ 0 \end{pmatrix}$$

where $y_1 \neq 0, y_2 \neq 0, \ldots, y_r \neq 0$. There r vectors are obviously linearly independent. Let $\mathbf{H} = [\mathbf{Y}_1 \mathbf{Y}_2 \ldots \mathbf{Y}_r]$ so that the rank of $\mathbf{H}$ is r. Then,

$$\mathbf{CH} = [\mathbf{CY}_1 \mathbf{CY}_2 \ldots \mathbf{CY}_r] = [\mathbf{X}_1 \mathbf{X}_2 \ldots \mathbf{X}_r] = \mathbf{G}$$

where $\mathbf{X}_1, \mathbf{X}_2, \ldots, \mathbf{X}_r$ are the latent vectors of $\mathbf{A}$ corresponding to $\lambda_1, \lambda_2, \ldots, \lambda_r$. Since $\mathbf{C}^{-1}$ exists, the rank of $\mathbf{G}$ is r and the theorem is now proved.

Theorem 1.16

If the latent roots of $\mathbf{A}$ are all distinct, then the n latent vectors are all linearly independent.

Proof

This follows immediately from theorem 1.15.

Theorem 1.17

If $\mathbf{A}$ has n linearly independent latent vectors then it is similar to a diagonal matrix.

Proof

Consider a matrix $\mathbf{C}$ whose columns are the latent vectors of $\mathbf{A}$. Then

$$\begin{aligned} \mathbf{AC} &= \mathbf{A}(\mathbf{X}_1 \mathbf{X}_2 \ldots \mathbf{X}_n) = (\lambda_1 \mathbf{X}_1 \lambda_2 \mathbf{X}_2 \ldots \lambda_n \mathbf{X}_n) \\ &= (\mathbf{X}_1 \mathbf{X}_2 \ldots \mathbf{X}_n) \begin{pmatrix} \lambda_1 & 0 & \ldots & 0 \\ 0 & \lambda_2 & \ldots & 0 \\ \vdots & \vdots & & \vdots \\ 0 & 0 & \ldots & \lambda_n \end{pmatrix} \\ &= \mathbf{CB} \end{aligned}$$

where $\mathbf{B}$ is diagonal. Since all the latent vectors $\mathbf{X}_1, \mathbf{X}_2, \ldots, \mathbf{X}_n$ are linearly independent, $\mathbf{C}^{-1}$ must exist. Hence

$$\mathbf{B} = \mathbf{C}^{-1} \mathbf{AC}$$

and the theorem is proved.

Theorem 1.18

If λ_i and λ_j are distinct latent roots of $\mathbf{A}$ with

$$\mathbf{AX} = \lambda_i \mathbf{X} \quad \text{and} \quad \mathbf{A}^T \mathbf{Y} = \lambda_j \mathbf{Y}$$

then $\mathbf{X}^T \mathbf{Y} = 0$, i.e. $\mathbf{X}$ and $\mathbf{Y}$ are bi-orthogonal.

Proof

From $\mathbf{AX} = \lambda_i \mathbf{X}$ we get

$$\lambda_i \mathbf{Y}^T \mathbf{X} = \mathbf{Y}^T \mathbf{AX} = (\mathbf{A}^T \mathbf{Y})^T \mathbf{X} = (\lambda_j \mathbf{Y})^T \mathbf{X} = \lambda_j \mathbf{Y}^T \mathbf{X}$$

Since $\lambda_i \neq \lambda_j$ it must be that $\mathbf{Y}^T \mathbf{X} = 0$, and hence $\mathbf{X}^T \mathbf{Y} = 0$.

Theorem 1.19

If $\mathbf{X}$ is a latent vector of $\mathbf{A}$ corresponding to the latent root $a+bi$, then $\mathbf{Y}$ is the latent vector corresponding to the latent root $a-bi$, where the elements of $\mathbf{Y}$ are the complex conjugates of those of $\mathbf{X}$.

Proof

Let us write $\mathbf{X} = \mathbf{X}_1 + i\mathbf{X}_2$ and $\mathbf{Y} = \mathbf{X}_1 - i\mathbf{X}_2$. We have

$$\mathbf{A}(\mathbf{X}_1 + i\mathbf{X}_2) = (a+bi)(\mathbf{X}_1 + i\mathbf{X}_2)$$

so that

$$\mathbf{AX}_1 + i\mathbf{AX}_2 = a\mathbf{X}_1 - b\mathbf{X}_2 + ia\mathbf{X}_2 + ib\mathbf{X}_1$$

Equating real and imaginary parts this gives

$$\mathbf{AX}_1 = a\mathbf{X}_1 - b\mathbf{X}_2 \tag{1.4}$$

and

$$\mathbf{AX}_2 = a\mathbf{X}_2 + b\mathbf{X}_1 \tag{1.5}$$

Multiplying (1.5) by $(-i)$ and adding to (1.4) we get

$$\mathbf{AX}_1 - i\mathbf{AX}_2 = a\mathbf{X}_1 - b\mathbf{X}_2 - ia\mathbf{X}_2 - ib\mathbf{X}_1$$

so that

$$\begin{aligned}\mathbf{A}(\mathbf{X}_1 - i\mathbf{X}_2) &= a\mathbf{X}_1 - ia\mathbf{X}_2 - ib\mathbf{X}_1 - b\mathbf{X}_2 \\ &= (a-bi)(\mathbf{X}_1 - i\mathbf{X}_2)\end{aligned}$$

or

$$\mathbf{AY} = (a-bi)\mathbf{Y}$$

as required.

1.5 Theorems Concerning Symmetric Matrices

Theorem 1.20

The latent roots and vectors of a real symmetric matrix are all real.

Proof

From theorem 1.5 we can put

$$\mathbf{B} = \mathbf{C}^{-1}\mathbf{AC}$$

where $\mathbf{B}$ is triangular and $\mathbf{C}$ is unitary, that is $\mathbf{C}^* = \mathbf{C}^{-1}$. Hence

$$\mathbf{B}^* = (\mathbf{C}^{-1}\mathbf{AC})^* = (\mathbf{C}^*\mathbf{AC})^* = \mathbf{C}^*\mathbf{A}^*\mathbf{C} = \mathbf{C}^{-1}\mathbf{AC} = \mathbf{B}$$

But we can only have $\mathbf{B}^* = \mathbf{B}$ if $\mathbf{B}$ has real diagonal elements and since the diagonal elements of $\mathbf{B}$ are the latent roots of $\mathbf{A}$, these roots must be real. It follows that the latent vectors of $\mathbf{A}$ must also be real.

Theorem 1.21

Every symmetric matrix is similar to a diagonal matrix.

Proof

From the above theorem we had $\mathbf{B}^* = \mathbf{B}$. Since $\mathbf{B}$ is triangular it must also be diagonal.

Theorem 1.22

A symmetric matrix has n linearly independent and orthogonal latent vectors.

Proof

From theorem 1.21 we can put

$$\mathbf{C}^{-1}\mathbf{A}\mathbf{C} = \mathbf{B} = \begin{pmatrix} \lambda_1 & 0 & \ldots & 0 \\ 0 & \lambda_2 & \ldots & 0 \\ \vdots & \vdots & & \vdots \\ 0 & 0 & \ldots & \lambda_n \end{pmatrix}$$

If the columns of $\mathbf{C}$ are $\mathbf{X}_1, \mathbf{X}_2, \ldots, \mathbf{X}_n$, this gives

$$\begin{aligned} \mathbf{AC} &= \mathbf{A}(\mathbf{X}_1\,\mathbf{X}_2\ldots\mathbf{X}_n) = (\mathbf{AX}_1\,\mathbf{AX}_2\ldots\mathbf{AX}_n) \\ &= \mathbf{CB} = (\lambda_1\mathbf{X}_1\,\lambda_2\mathbf{X}_2\ldots\lambda_n\mathbf{X}_n) \end{aligned}$$

and it follows that $\mathbf{X}_1, \mathbf{X}_2, \ldots, \mathbf{X}_n$ are all latent vectors of $\mathbf{A}$. Since $\mathbf{C}^{-1}$ exists they must all be linearly independent.

Furthermore, since from theorem 1.20 the latent roots and hence the latent vectors of $\mathbf{A}$ are real we have

$$\mathbf{C}^{-1} = \mathbf{C}^* = \mathbf{C}^T$$

so that

$$(\mathbf{X}_1\,\mathbf{X}_2\ldots\mathbf{X}_n)^T(\mathbf{X}_1\,\mathbf{X}_2\ldots\mathbf{X}_n) = \mathbf{I}$$

which means that

$$\mathbf{X}_i^T\mathbf{X}_j = 0 \quad \text{whenever } i \neq j$$

This completes the proof of the theorem.

Further results and theorems will be given as and when they are required.

In the next section just a few of the applications of latent roots and vectors are briefly discussed.

1.6 Exercises

Section 1.1

1.1. Find the latent roots and vectors of the following matrices

(i) $\begin{pmatrix} 7 & 5 \\ -4 & -2 \end{pmatrix}$ (ii) $\begin{pmatrix} a & -b \\ b & a \end{pmatrix}$

(iii) $\begin{pmatrix} -1 & -1 & 5 \\ 3 & 1 & -1 \\ -2 & -1 & 6 \end{pmatrix}$ (iv) $\begin{pmatrix} 3 & 5 & -2 \\ -3 & 1 & 3 \\ 2 & 5 & -1 \end{pmatrix}$

1.2. Show that all vectors are latent vectors of the unit matrix.

Section 1.2

1.3. Show that the matrix $\mathbf{R}$ of § 1.2 has the effect of rotating the x, y-axes through an angle $-\theta$. (Alternatively we can think of $\mathbf{R}$ as rotating a point through an angle θ.)

1.4. Use the matrix of rotation $\mathbf{R}$ to show that

$$\sin(\alpha+\theta) = \sin\alpha\cos\theta+\cos\alpha\sin\theta$$

Show that

$$\mathbf{R}^n = \begin{pmatrix} \cos n\theta & -\sin n\theta \\ \sin n\theta & \cos n\theta \end{pmatrix}$$

Show also that $\mathbf{R}$ is isomorphic to the complex number $\cos\theta+i\sin\theta$. Hence prove De Moivre's theorem.

1.5. Show that the matrix

$$\mathbf{H} = \begin{pmatrix} \cosh\theta & \sinh\theta \\ \sinh\theta & \cosh\theta \end{pmatrix}$$

has the effect of rotating a point (x, y) on the hyperbola $x^2-y^2 = r^2$ through a 'hyperbolic angle' θ. Hence prove that

$$\sinh(\alpha+\theta) = \sinh\alpha\cosh\theta+\cosh\alpha\sinh\theta$$

1.6. If $\lambda \neq 0$ is a latent root of $\mathbf{AB}$, show that it is also a latent root of $\mathbf{BA}$. What is the connection between their corresponding latent vectors? Deduce that if $\mathbf{AB} = \mathbf{BA}$ then $\mathbf{A}$ and $\mathbf{B}$ have a common latent vector.

1.7. Show that a matrix $\mathbf{A} \neq \mathbf{I}$ cannot be similar to $\mathbf{I}$. Hence give an example of two matrices with the same characteristic equation that are not similar.

1.8. Show that a unitary matrix $\mathbf{C}$ can be found with a given first column $\mathbf{X}$ such that $\mathbf{X}^*\mathbf{X} = 1$.

1.9. If $\mathbf{A}^{-1}$ and $\mathbf{B}^{-1}$ both exist show that $(\mathbf{AB})^{-1} = \mathbf{B}^{-1}\mathbf{A}^{-1}$.

Section 1.3

1.10. If $\mathbf{A}$ is unitary show that $|\lambda| = 1$ for all λ.

1.11. By finding the latent roots of the matrix $\mathbf{R}$ of question 1.4 give an alternative proof for De Moivre's theorem when n is an integer.

1.12. Prove that $\mathbf{A}^m$ remains finite as $m \to \infty$ if, and only if, $|\lambda| \leqslant 1$ for all λ.

1.13. The minimal polynomial of a matrix $\mathbf{A}$ is defined to be the polynomial $h(\lambda)$ of least degree such that $h(\mathbf{A}) = 0$. Show that $h(\lambda)$ is a factor of the characteristic equation of $\mathbf{A}$. Show also that any latent root of $\mathbf{A}$ is a root of $h(\lambda)$.

1.14. Prove that every latent root of $\mathbf{A}$ lies on or inside at least one of the circles.

$$|\lambda - a_{ii}| = \sum_{j=1}^{n} |a_{ij}| - |a_{ii}|$$

This is known as *Brauer's theorem.*

Section 1.4

1.15. Prove that if $\mathbf{A}$ is similar to a diagonal matrix then it has n linearly independent latent vectors.

Section 1.5

1.16. Show that $(\mathbf{AB})^* = \mathbf{B}^*\mathbf{A}^*$.

1.17. A matrix such that $\mathbf{A} = \mathbf{A}^*$ is called *Hermitian.* Prove that the latent roots of a Hermitian matrix are all real. Also prove that a Hermitian matrix is similar to a diagonal matrix.

Miscellaneous

1.18. Find the latent roots of the magic squares

$$\text{(i)}\quad \mathbf{A} = \begin{pmatrix} 8 & 1 & 6 \\ 3 & 5 & 7 \\ 4 & 9 & 2 \end{pmatrix}; \quad \text{(ii)}\quad \mathbf{A} = \begin{pmatrix} 16 & 2 & 3 & 13 \\ 5 & 11 & 10 & 8 \\ 9 & 7 & 6 & 12 \\ 4 & 14 & 15 & 1 \end{pmatrix}$$

($\mathbf{A}$ is magic if the elements in each row, column and the two main diagonals add up to the same number, this number being called the *magic number.*)

Show that one latent root of a magic square is the magic number.

1.19. Show that the characteristic equation of the $n \times n$ matrix,

$$\mathbf{A} = \begin{pmatrix} 0 & 1 & 0 & 0 & \dots & 0 & 0 & 0 \\ 1 & 0 & 1 & 0 & \dots & 0 & 0 & 0 \\ 0 & 1 & 0 & 1 & \dots & 0 & 0 & 0 \\ \vdots & \vdots & \vdots & \vdots & & \vdots & \vdots & \vdots \\ 0 & 0 & 0 & 0 & \dots & 0 & 1 & 0 \\ 0 & 0 & 0 & 0 & \dots & 1 & 0 & 1 \\ 0 & 0 & 0 & 0 & \dots & 0 & 1 & 0 \end{pmatrix}$$

is given by

$$f(\lambda) = \begin{cases} \lambda^n - {}^{n-1}C_1\lambda^{n-2} + {}^{n-2}C_2\lambda^{n-4} - {}^{n-3}C_3\lambda^{n-6} + \ldots + (-1)^{n/2} = 0, & n \text{ even} \\ -\lambda^n + {}^{n-1}C_1\lambda^{n-2} - {}^{n-2}C_2\lambda^{n-4} + {}^{n-3}C_3\lambda^{n-6} + \ldots \\ \qquad + (-1)^{(n+1)/2}[(n+1)/2]\lambda = 0, & n \text{ odd} \end{cases}$$

1.20. If

$$\mathbf{A} = \begin{pmatrix} 0 & a & a & \ldots & a \\ a & 0 & a & \ldots & a \\ a & a & 0 & \ldots & a \\ \vdots & \vdots & \vdots & & \vdots \\ a & a & a & \ldots & 0 \end{pmatrix}$$

show that the characteristic equation of **A** is

$$(-1)^{n-1}(a+\lambda)^{n-1}(na-a-\lambda) = 0$$

where n is the order of the matrix **A**. Find the latent vectors of **A**.

1.21. If

$$\mathbf{A} = \begin{pmatrix} a_1^2 & a_1a_2 & a_1a_3 & \ldots & a_1a_n \\ a_2a_1 & a_2^2 & a_2a_3 & \ldots & a_2a_n \\ a_3a_1 & a_3a_2 & a_3^2 & \ldots & a_3a_n \\ \vdots & \vdots & \vdots & & \vdots \\ a_na_1 & a_na_2 & a_na_3 & \ldots & a_n^2 \end{pmatrix}$$

show that the only latent roots of **A** are

$$\lambda = 0 \quad \text{and} \quad \lambda = \sum_{i=1}^{n} a_i^2$$

1.22. If

$$\mathbf{A} = \begin{pmatrix} 0 & 0 & 0 & \ldots & 0 & a_1 \\ 0 & 0 & 0 & \ldots & 0 & a_2 \\ \vdots & \vdots & \vdots & & \vdots & \vdots \\ 0 & 0 & 0 & \ldots & 0 & a_{n-1} \\ b_1 & b_2 & b_3 & \ldots & b_{n-1} & a_nb_n \end{pmatrix}$$

show that the characteristic equation of **A** is

$$(-\lambda)^{n-2}\left(\lambda^2 - a_nb_n\lambda - \sum_{i=1}^{n-1} a_ib_i\right) = 0$$

1.23. If **A** is the circulant given by

$$\mathbf{A} = \begin{pmatrix} a_1 & a_2 & a_3 & \cdots & a_n \\ a_n & a_1 & a_2 & \cdots & a_{n-1} \\ a_{n-1} & a_n & a_1 & \cdots & a_{n-2} \\ \vdots & \vdots & \vdots & & \vdots \\ a_2 & a_3 & a_4 & \cdots & a_1 \end{pmatrix}$$

show that

$$|\mathbf{A}-\lambda\mathbf{I}| = f(e_0)f(e_1)f(e_2)\ldots f(e_{n-1})$$

where

$$f(x) = a_1+a_2x+a_3x^2+\ldots+a_nx^{n-1}-\lambda$$

and

$$e_k = \mathrm{e}^{(2\pi k/n)}$$

That is, $e_0, e_1, \ldots, e_{n-1}$ are the n roots of $\sqrt[n]{1}$.

1.24. A symmetric matrix **A** is said to be positive definite if $\mathbf{X}^T\mathbf{A}\mathbf{X} > 0$ for every non-zero real vector **X**. Show that **A** is positive definite if all its latent roots are positive.

1.25. A matrix is said to be normal if $\mathbf{A}\mathbf{A}^* = \mathbf{A}^*\mathbf{A}$. Prove that

(i) The latent vectors corresponding to distinct latent roots of a normal matrix are orthogonal.

(ii) If **A** is normal then **A** and $\mathbf{A}^*$ have the same latent vectors.

1.26. If $\mathbf{A}^r = 0$ for some positive integer r then **A** is said to be nilpotent. Prove that the latent roots of a nilpotent matrix are all zero.

1.27. If $\mathbf{A}^2 = \mathbf{A}$ then **A** is said to be idempotent. Show that the latent roots of an idempotent matrix are all zero or unity.

1.28. Show that **AB** and **BA** have the same characteristic equation. (Question 1.6 had the restriction that $\lambda \neq 0$.)

Chapter 2

APPLICATIONS OF LATENT ROOTS AND LATENT VECTORS

The first application given is useful in that it helps give a geometric understanding to the latent root and vector problem.

2.1 Axes of Symmetry of a Conic Section

The general two-dimensional conic whose centre is at the origin is given by

$$f(x, y) = ax^2 + 2hxy + by^2 = 1$$

We can write this in matrix form as

$$\begin{pmatrix} x & y \end{pmatrix} \begin{pmatrix} a & h \\ h & b \end{pmatrix} \begin{pmatrix} x \\ y \end{pmatrix} = \begin{pmatrix} 1 \end{pmatrix} \tag{2.1}$$

or

$$\mathbf{X}^T \mathbf{A} \mathbf{X} = 1 \tag{2.2}$$

The slope of the normal at a point $P(x_1, y_1)$ on the curve is given by

$$S_n = \frac{\partial f}{\partial y} \Big/ \frac{\partial f}{\partial x} = \frac{2hx_1 + 2by_1}{2ax_1 + 2hy_1} = \frac{hx_1 + by_1}{ax_1 + hy_1}$$

The normal will be an axis of symmetry of the conic if its slope is equal to the slope of the line OP, O being the origin. If this is the case,

$$\frac{hx_1 + by_1}{ax_1 + hy_1} = \frac{y_1}{x_1}$$

This will be true if there exists λ such that

$$ax_1 + hy_1 = \lambda x_1$$

and

$$hx_1 + by_1 = \lambda y_1$$

that is, if

$$\begin{pmatrix} a & h \\ h & b \end{pmatrix} \begin{pmatrix} x_1 \\ y_1 \end{pmatrix} = \lambda \begin{pmatrix} x_1 \\ y_1 \end{pmatrix}$$

or

$$\mathbf{AX} = \lambda \mathbf{X}$$

Clearly any vector, **X**, satisfying this equation will be an axis of symmetry of the conic. From theorem 1.22 we know that there will be two such vectors $\mathbf{X}_1$ and $\mathbf{X}_2$, and that $\mathbf{X}_1^T\mathbf{X}_2 = 0$.

Furthermore, if **X** is a latent vector of **A**, from equation (2.2) we get

$$\mathbf{X}^T\mathbf{A}\mathbf{X} = \mathbf{X}^T\lambda\mathbf{X} = \lambda\mathbf{X}^T\mathbf{X} = 1$$

but

$$\mathbf{X}^T\mathbf{X} = (x^2+y^2) = (r^2)$$

where r is the distance of P from the origin. Hence,

$$r^2 = 1/\lambda$$

This also helps us to rotate the conic so that its axes lie along the x, y-axes. We wish to rotate the axes of symmetry, say x' and y', through an angle $-\theta$ so that they lie along the x, y-axes. To achieve this we put (see § 1.2),

$$\begin{pmatrix} x \\ y \end{pmatrix} = \begin{pmatrix} \cos\theta & -\sin\theta \\ \sin\theta & \cos\theta \end{pmatrix} \begin{pmatrix} x' \\ y' \end{pmatrix} = \mathbf{R}\mathbf{Y} \tag{2.3}$$

which also gives

$$(\,x \quad y\,) = (x' \quad y'\,) \begin{pmatrix} \cos\theta & \sin\theta \\ -\sin\theta & \cos\theta \end{pmatrix} = \mathbf{Y}^T\mathbf{R}^{-1} \tag{2.4}$$

Now notice that the point P is given by

$$x_1 = r\cos\theta \quad \text{and} \quad y_1 = r\sin\theta$$

and if Q is the point (x_2, y_2) lying on the intersection of the curve and the other latent vector, then clearly

$$x_2 = -r'\sin\theta \quad \text{and} \quad y_2 = r'\cos\theta$$

and hence the columns of **R** are latent vectors of **A**. So substituting (2.3) and (2.4) into (2.1) we get

$$\mathbf{X}^T\mathbf{A}\mathbf{X} = \mathbf{Y}^T\mathbf{R}^{-1}\mathbf{A}\mathbf{R}\mathbf{Y} = \mathbf{Y}^T\mathbf{B}\mathbf{Y}$$

where, from theorems 1.22 and 1.23, **B** is diagonal with the latent roots of **A** as its leading diagonal.

Hence the equation of the conic becomes

$$(\,x' \quad y'\,) \begin{pmatrix} \lambda_1 & 0 \\ 0 & \lambda_2 \end{pmatrix} \begin{pmatrix} x' \\ y' \end{pmatrix} = (\,1\,)$$

or

$$\lambda_1 x^2 + \lambda_2 y^2 = 1$$

We can see that a knowledge of the latent roots and vectors of **A** is extremely useful for investigating the conic section. These results are easily extended to higher dimensions and to conics whose centres are not at the origin.

Example 2.1

Take as an example the ellipse given by

$$8x^2 - 4xy + 5y^2 = 1 \quad \text{(See Fig. 3.)}$$

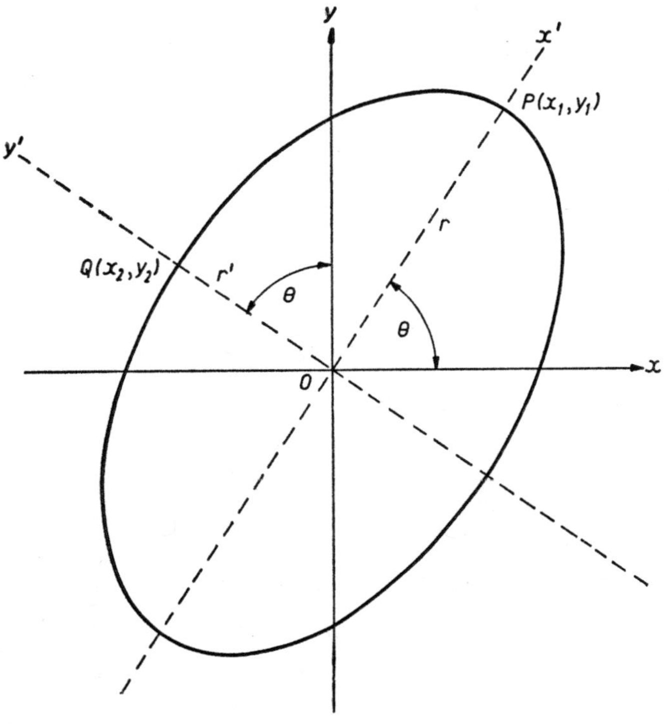

FIG. 3

In matrix form the equation of the ellipse is

$$(x \quad y)\begin{pmatrix} 8 & -2 \\ -2 & 5 \end{pmatrix}\begin{pmatrix} x \\ y \end{pmatrix} = (1)$$

or

$$\mathbf{X}^T\mathbf{A}\mathbf{X} = 1$$

The characteristic equation of **A** is

$$\begin{vmatrix} 8-\lambda & -2 \\ -2 & 5-\lambda \end{vmatrix} = 0$$

which gives

$$\lambda^2 - 13\lambda + 36 = 0$$

or

$$(\lambda - 9)(\lambda - 4) = 0$$

so that

$$\lambda_1 = 9 \quad \text{and} \quad \lambda_2 = 4$$

Using $\mathbf{AX} = \lambda\mathbf{X}$ we have

$$8x - 2y = \lambda x$$

and

$$-2x + 5y = \lambda y$$

When $\lambda = 9$,

$$-x = 2y$$

$$-2x = 4y$$

so that

$$y = -\tfrac{1}{2}x$$

When $\lambda = 4$,

$$4x = 2y$$

$$-2x = -y$$

so that

$$y = 2x$$

The major axis of the ellipse is the line $y = 2x$, and its length is $r = 1/\sqrt{\lambda_2} = \frac{1}{2}$. The minor axis of the ellipse is the line $y = -\frac{1}{2}x$, and its length is $r' = 1/\sqrt{\lambda_1} = \frac{1}{3}$. If we rotate the ellipse so that its axes lie along the x and y axes we get the equation

$$4x^2 + 9y^2 = 1$$

2.2 Jacobi and Gauss–Seidel Methods

Two important iterative methods of solving a set of simultaneous linear equations are the Jacobi and the Gauss–Seidel methods. Latent roots play an important role here in determining the convergence of these methods.

We shall first outline the two methods. We wish to solve the equations

$$\begin{matrix} a_{11}x_1 + a_{12}x_2 + \ldots + a_{1n}x_n = b_1 \\ a_{21}x_1 + a_{22}x_2 + \ldots + a_{2n}x_n = b_2 \\ \vdots \qquad \vdots \qquad \qquad \vdots \qquad \vdots \\ a_{n1}x_1 + a_{n2}x_2 + \ldots + a_{nn}x_n = b_n \end{matrix}$$

or as a matrix equation

$$\mathbf{AX} = \mathbf{B} \tag{2.5}$$

Assuming $a_{ii} \neq 0$, we rewrite the equations as

$$\begin{aligned}
x_1 &= 1/a_{11}\,(\qquad\qquad -a_{12}x_2 - \ldots - a_{1n}x_n) + b_1/a_{11} \\
x_2 &= 1/a_{22}\,(-a_{21}x_1 \qquad\qquad - \ldots - a_{2n}x_n) + b_2/a_{22} \\
&\vdots \\
x_n &= 1/a_{nn}(-a_{n1}x_1 - a_{n2}x_2 - \ldots \qquad\qquad) + b_n/a_{nn}
\end{aligned}$$

In the Jacobi method we take an initial approximation to the solution of the equations and substitute this into the right-hand side of the above equations to produce an improved solution. We then substitute this improved solution in the right-hand side of the equations, and so on.

If we denote x_{ir} as the rth approximation to x_i we can represent this process as

$$\begin{aligned}
x_{1,r+1} &= 1/a_{11}\,(\qquad\qquad -a_{12}x_{2r} - \ldots - a_{1n}x_{nr}) + b_1/a_{11} \\
x_{2,r+1} &= 1/a_{22}\,(-a_{21}x_{1r} \qquad\qquad - \ldots - a_{2n}x_{nr}) + b_2/a_{22} \\
&\vdots \\
x_{n,r+1} &= 1/a_{nn}(-a_{n1}x_{1r} - a_{n2}x_{2r} - \ldots \qquad\qquad) + b_n/a_{nn}
\end{aligned}$$

or in matrix form as

$$\mathbf{X}_{r+1} = \mathbf{P}\mathbf{X}_r + \mathbf{Q} \tag{2.6}$$

where

$$\mathbf{P} = \begin{pmatrix} 1 & 0 & \ldots & 0 \\ 0 & 1 & \ldots & 0 \\ \vdots & \vdots & & \vdots \\ 0 & 0 & \ldots & 1 \end{pmatrix} - \begin{pmatrix} 1/a_{11} & 0 & \ldots & 0 \\ 0 & 1/a_{22} & \ldots & 0 \\ \vdots & \vdots & & \vdots \\ 0 & 0 & \ldots & 1/a_{nn} \end{pmatrix} \begin{pmatrix} a_{11} & a_{12} & \ldots & a_{1n} \\ a_{21} & a_{22} & \ldots & a_{2n} \\ \vdots & \vdots & & \vdots \\ a_{n1} & a_{n2} & \ldots & a_{nn} \end{pmatrix}$$

$$= \mathbf{I} - \mathbf{D}^{-1}\mathbf{A} \tag{2.7}$$

and

$$\mathbf{Q} = \mathbf{D}^{-1}\mathbf{B}$$

The Gauss–Seidel method varies from Jacobi in that as soon as an approximation to x_i is found, it is used in all the remaining equations. We represent this as

$$\begin{aligned}
x_{1,r+1} &= 1/a_{11}\,(\qquad\qquad -a_{12}x_{2r} - \ldots - a_{1n}x_{nr}) + b_1/a_{11} \\
x_{2,r+1} &= 1/a_{22}\,(-a_{21}x_{1,r+1} \qquad\qquad - \ldots - a_{2n}x_{nr}) + b_2/a_{22} \\
&\vdots \\
x_{n,r+1} &= 1/a_{nn}(-a_{n1}x_{1,r+1} - a_{n2}x_{2,r+1} - \ldots \qquad\qquad) + b_n/a_{nn}
\end{aligned}$$

or in matrix form as

$$\mathbf{D}^{-1}(\mathbf{L}+\mathbf{D})\mathbf{X}_{r+1} = -\mathbf{D}^{-1}\mathbf{U}\mathbf{X}_r+\mathbf{Q}$$

where **L** is a lower triangular matrix with zeros in each position of its leading diagonal (hence $\{\mathbf{D}^{-1}(\mathbf{L}+\mathbf{D})\}^{-1}$ exists) and **U** an upper triangular matrix with zeros in each position of its leading diagonal, and

$$\mathbf{D}+\mathbf{L}+\mathbf{U} = \mathbf{A} \tag{2.8}$$

Since $\{\mathbf{D}^{-1}(\mathbf{L}+\mathbf{D})\}^{-1}$ exists, we have

$$\mathbf{X}_{r+1} = -(\mathbf{L}+\mathbf{D})^{-1}\mathbf{U}\mathbf{X}_r+(\mathbf{L}+\mathbf{D})^{-1}\mathbf{B} \tag{2.9}$$

which is of the same form as equation (2.6). For this reason we need only investigate an equation of the form

$$\mathbf{X}_{r+1} = \mathbf{M}\mathbf{X}_r+\mathbf{Y}$$

which gives

$$\mathbf{X}_1 = \mathbf{M}\mathbf{X}_0+\mathbf{Y}$$

$$\mathbf{X}_2 = \mathbf{M}\mathbf{X}_1+\mathbf{Y} = \mathbf{M}(\mathbf{M}\mathbf{X}_0+\mathbf{Y})+\mathbf{Y} = \mathbf{M}^2\mathbf{X}_0+\mathbf{M}\mathbf{Y}+\mathbf{Y}$$

$$\mathbf{X}_3 = \mathbf{M}\mathbf{X}_2+\mathbf{Y} = \mathbf{M}(\mathbf{M}^2\mathbf{X}_0+\mathbf{M}\mathbf{Y}+\mathbf{Y})+\mathbf{Y} = \mathbf{M}^3\mathbf{X}_0+\mathbf{M}^2\mathbf{Y}+\mathbf{M}\mathbf{Y}+\mathbf{Y}$$

and it is a simple matter to show by induction that

$$\mathbf{X}_r = \mathbf{M}^r\mathbf{X}_0+\mathbf{M}^{r-1}\mathbf{Y}+\mathbf{M}^{r-2}\mathbf{Y}+\ldots+\mathbf{M}\mathbf{Y}+\mathbf{Y}$$

Premultiplying by **M** gives

$$\mathbf{M}\mathbf{X}_r = \mathbf{M}^{r+1}\mathbf{X}_0+\mathbf{M}^r\mathbf{Y}+\mathbf{M}^{r-1}\mathbf{Y}+\ldots+\mathbf{M}^2\mathbf{Y}+\mathbf{M}\mathbf{Y}$$

and subtracting the first of these from the second we get

$$\mathbf{M}\mathbf{X}_r-\mathbf{X}_r = \mathbf{M}^{r+1}\mathbf{X}_0-\mathbf{M}^r\mathbf{X}_0+\mathbf{M}^r\mathbf{Y}-\mathbf{Y}$$

so that

$$(\mathbf{M}-\mathbf{I})\mathbf{X}_r = \mathbf{M}^r(\mathbf{M}-\mathbf{I})\mathbf{X}_0+(\mathbf{M}^r-\mathbf{I})\mathbf{Y}$$

Providing that **M** does not have a latent root equal to unity $(\mathbf{M}-\mathbf{I})^{-1}$ exists. So making this assumption

$$\mathbf{X}_r = (\mathbf{M}-\mathbf{I})^{-1}\mathbf{M}^r(\mathbf{M}-\mathbf{I})\mathbf{X}_0+(\mathbf{M}-\mathbf{I})^{-1}(\mathbf{M}^r-\mathbf{I})\mathbf{Y}$$

As $r\to\infty$ we obviously require $\mathbf{X}_r$ to converge to a finite limit independent of our initial value $\mathbf{X}_0$. This will be true if $\mathbf{M}^r\to 0$ as $r\to\infty$. If this is true

$$\lim_{r\to\infty}\mathbf{X}_r = \mathbf{X} = (\mathbf{M}-\mathbf{I})^{-1}(-\mathbf{I})\mathbf{Y}$$

which gives

$$(\mathbf{M}-\mathbf{I})\mathbf{X} = -\mathbf{I}\mathbf{Y}$$

or

$$\mathbf{X} = \mathbf{M}\mathbf{X}+\mathbf{Y}$$

which clearly satisfies our initial equations.

From theorem 1.7, $\lim_{r\to\infty} \mathbf{M}^r = 0$ if and only if $|\alpha| < 1$ for all α, where α is a latent root of $\mathbf{M}$. (If this is the case the above assumption that there is no $\alpha = 1$ is justified.) Let us try to translate this result back to the original problems.

We take first the Jacobi method. From equation (2.7) we see that

$$\mathbf{M} = \mathbf{I} - \mathbf{D}^{-1}\mathbf{A}$$

so that if the method is to converge we require $|1-\lambda| < 1$ for all λ, where λ is a latent root of $\mathbf{D}^{-1}\mathbf{A}$.

From the form of $\mathbf{I} - \mathbf{D}^{-1}\mathbf{A}$ and using Gerschgorin's theorem (theorem 1.10) we get as a sufficient condition for convergence of the Jacobi method that

$$\sum_{j=1}^{n} |a_{ij}| < 2|a_{ii}| \quad \text{for all } i \tag{2.10}$$

Taking the Gauss–Seidel method we have from equation (2.9) that

$$-(\mathbf{L}+\mathbf{D})^{-1}\mathbf{U} = -(\mathbf{L}+\mathbf{D})^{-1}(\mathbf{A}-\mathbf{L}-\mathbf{D}) = (\mathbf{L}+\mathbf{D})^{-1}(\mathbf{L}+\mathbf{D}-\mathbf{A})$$
$$= \mathbf{I} - (\mathbf{L}+\mathbf{D})^{-1}\mathbf{A}$$

which gives that the necessary condition for the process to converge is given by $|1-\beta| < 1$ for all β, where β is a latent root of $(\mathbf{L}+\mathbf{D})^{-1}\mathbf{A}$.

It can be shown that a sufficient condition for the Gauss–Seidel method to converge is the same as the condition given above for the Jacobi method.†

Since (2.10) gives a sufficient condition for the convergence of both the Jacobi and the Gauss–Seidel methods, we can see why it is generally recommended that we arrange our set of equations in order that they have a strong leading diagonal.‡

Although the given sufficient conditions are the same for both methods, it is clear that the necessary conditions are not the same. See exercise 2.7 for examples of each of the cases where one method converges and the other diverges. Varga has discussed the rate of convergence of the two methods.§

2.3 Stability of the Numerical Solution of Partial Differential Equations

In stability we are concerned with the propagation of errors in the numerical solution of a problem. If the errors decay as we proceed the method is said to be *stable*, otherwise it is *unstable*. Instability is usually caused by the growth of rounding errors or by the presence of an unwanted (parasitic)

† See reference 1, p. 73.
‡ See reference 2, p. 73.
§ See reference 1.

solution. An investigation of the stability of a method can often involve the finding of the latent roots of a matrix. We take as an example the solution of the parabolic partial differential equation given by

$$\frac{\partial u}{\partial t} = \frac{\partial^2 u}{\partial x^2} \tag{2.11}$$

We wish to find values of $u(x,t)$ for given values of x and t (Fig. 4).

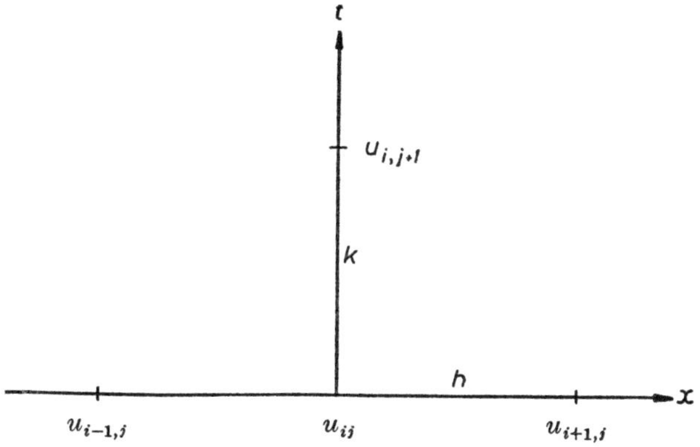

FIG. 4

Suppose that we know the value of $u(i,j) = u_{ij}$. Then by Taylor's series we have

$$u(x+h,t) = u_{i+1,j} = u_{ij} + h.\frac{\partial u}{\partial x} + \frac{h^2}{2!}.\frac{\partial^2 u}{\partial x^2} + \ldots \tag{2.12}$$

$$u(x-h,t) = u_{i-1,j} = u_{ij} - h.\frac{\partial u}{\partial x} + \frac{h^2}{2!}.\frac{\partial^2 u}{\partial x^2} - \ldots \tag{2.13}$$

$$u(x,t+k) = u_{i,j+1} = u_{ij} + k.\frac{\partial u}{\partial t} + \frac{k^2}{2!}.\frac{\partial^2 u}{\partial t^2} + \ldots \tag{2.14}$$

Adding equations (2.12) and (2.13) and ignoring terms in h^4 and higher we get

$$u_{i+1,j} + u_{i-1,j} \simeq 2u_{ij} + h^2.\frac{\partial^2 u}{\partial x^2}$$

so that

$$\frac{\partial^2 u}{\partial x^2} \simeq \frac{u_{i-1,j} - 2u_{ij} + u_{i+1,j}}{h^2} \tag{2.15}$$

From equation (2.14) if we ignore terms in k^2 and higher we get

$$u_{i,j+1} \simeq u_{ij} + k \cdot \frac{\partial u}{\partial t}$$

so that

$$\frac{\partial u}{\partial t} \simeq \frac{u_{i,j+1} - u_{ij}}{k} \tag{2.16}$$

From equation (2.11) we can equate equations (2.15) and (2.16) to give a finite-difference approximation to the problem. I intend to look instead at the more interesting case of the Crank–Nicolson method. They replaced equation (2.15) by

$$\frac{\partial^2 u}{\partial x^2} \simeq \frac{1}{2}\left(\frac{u_{i-1,j} - 2u_{ij} + u_{i+1,j}}{h^2} + \frac{u_{i-1,j+1} - 2u_{i,j+1} + u_{i+1,j+1}}{h^2}\right)$$

which is the mean of (2.15) as it stands and (2.15) with $j+1$ instead of j. So equating this to equation (2.16) and putting $r = k/h^2$ we get

$$u_{i,j+1} - u_{ij} = r/2(u_{i-1,j} - 2u_{ij} + u_{i+1,j} + u_{i-1,j+1} - 2u_{i,j+1} + u_{i+1,j+1})$$

which gives

$$-ru_{i-1,j+1} + (2+2r)\,u_{i,j+1} - ru_{i+1,j+1} = ru_{i-1,j} + (2-2r)\,u_{ij} + ru_{i+1,j} \tag{2.17}$$

If we know the initial and boundary values for $j = 0$ and $i = 1, 2, \ldots, n$, then from equation (2.17) we can obtain n simultaneous linear equations for the n values when $j = 1$. Having found the values for $j = 1$ we can then repeat the process to find the n values when $j = 2$, and so on.† The equations arising from equation (2.17) for $i = 1, 2, \ldots, n$ can be written in matrix form, assuming $u_{0,j} = u_{0,j+1} = u_{n+1,j} = u_{n+1,j+1} = 0$:

$$\begin{pmatrix} (2+2r) & -r & 0 & \ldots & 0 & 0 & 0 \\ -r & (2+2r) & -r & \ldots & 0 & 0 & 0 \\ \vdots & \vdots & \vdots & & \vdots & \vdots & \vdots \\ 0 & 0 & 0 & \ldots & -r & (2+2r) & -r \\ 0 & 0 & 0 & \ldots & 0 & -r & (2+2r) \end{pmatrix} \begin{pmatrix} u_{1,j+1} \\ u_{2,j+1} \\ \vdots \\ u_{n-1,j+1} \\ u_{n,j+1} \end{pmatrix}$$

$$= \begin{pmatrix} (2-2r) & r & 0 & \ldots & 0 & 0 & 0 \\ r & (2-2r) & r & \ldots & 0 & 0 & 0 \\ \vdots & \vdots & \vdots & & \vdots & \vdots & \vdots \\ 0 & 0 & 0 & \ldots & r & (2-2r) & r \\ 0 & 0 & 0 & \ldots & 0 & r & (2-2r) \end{pmatrix} \begin{pmatrix} u_{1j} \\ u_{2j} \\ \vdots \\ u_{n-1,j} \\ u_{nj} \end{pmatrix}$$

† See reference 4, p. 18, for a worked example.

or

$$\mathbf{BU}_{j+1} = (4\mathbf{I}-\mathbf{B})\,\mathbf{U}_j$$

which gives

$$\mathbf{U}_{j+1} = (4\mathbf{B}^{-1}-\mathbf{I})\,\mathbf{U}_j = \mathbf{AU}_j \tag{2.18}$$

If our starting vector is $\mathbf{U}_0$ we have

$$\begin{aligned}\mathbf{U}_1 &= \mathbf{AU}_0 & & \\ \mathbf{U}_2 &= \mathbf{AU}_1 &&= \mathbf{A}^2\mathbf{U}_0 \\ \vdots & \quad\ \vdots && \quad\ \vdots \\ \mathbf{U}_m &= \mathbf{AU}_{m-1} &&= \mathbf{A}^m\mathbf{U}_0\end{aligned}$$

Suppose that instead of starting with the exact vector $\mathbf{U}_0$ we start with a vector $\mathbf{V}_0$, because our initial data have been rounded or have experimental errors, then instead of finding $\mathbf{U}_m$ we get $\mathbf{V}_m$, where

$$\mathbf{V}_m = \mathbf{A}^m\mathbf{V}_0$$

If we define $\mathbf{E}_i$ as the error vector due to rounding errors by

$$\mathbf{E}_i = \mathbf{U}_i - \mathbf{V}_i$$

we get

$$\mathbf{E}_m = \mathbf{U}_m - \mathbf{V}_m = \mathbf{A}^m\mathbf{U}_0 - \mathbf{A}^m\mathbf{V}_0 = \mathbf{A}^m(\mathbf{U}_0 - \mathbf{V}_0) = \mathbf{A}^m\mathbf{E}_0$$

If the method is to be stable we wish the rounding error to decay as we proceed, that is,

$$\mathbf{E}_m \to 0 \quad \text{as} \quad m \to \infty$$

From theorem 1.7 this will be true if $|\lambda| < 1$ for all λ where λ is a latent root of $\mathbf{A}$. Equation (2.18) gives

$$\mathbf{A} = 4\mathbf{B}^{-1} - \mathbf{I}$$

so that using theorem 1.13 we get

$$\lambda = \frac{4}{\beta} - 1$$

where β is a latent root of $\mathbf{B}$. Hence, we require

$$\left| \frac{4}{\beta} - 1 \right| < 1 \tag{2.19}$$

Now $\mathbf{B}$ is a common tridiagonal matrix so that its latent roots are given by †

$$\beta_m = (2+2r) + 2\sqrt{(r^2)}\cos\frac{m\pi}{n+1}, \quad m = 1, 2, \ldots, n$$

† See Appendix 1.

Since $|\cos\theta| \leqslant 1$ we clearly have

$$2 < \beta \leqslant 2 + 4r$$

$2 < \beta$ guarantees the satisfaction of condition (2.19). We reach the interesting conclusion that the Crank–Nicolson method is stable for any choice of $r > 0$.

It is hoped that this example has demonstrated the importance of latent roots in this field.

2.4 Simultaneous Differential Equations

Latent roots and vectors play an extremely important part in the solution of simultaneous differential equations. Only an elementary introduction to the case of first-order linear equations with constant coefficients is given here.

The single first-order equation

$$\frac{dx}{dt} = ax$$

has the general solution

$$x = ke^{at}$$

where k is a constant. Suppose that we have the equations

$$\frac{dx_1}{dt} = ax_1 + bx_2$$

$$\frac{dx_2}{dt} = cx_1 + dx_2$$

These can be written as

$$\frac{d}{dt}\begin{pmatrix} x_1 \\ x_2 \end{pmatrix} = \begin{pmatrix} a & b \\ c & d \end{pmatrix} \begin{pmatrix} x_1 \\ x_2 \end{pmatrix}$$

or

$$\frac{d\mathbf{X}}{dt} = \mathbf{AX} \qquad (2.20)$$

Suppose that the solution is of the form

$$\begin{aligned} x_1 &= k_1 e^{\lambda t} \\ x_2 &= k_2 e^{\lambda t} \end{aligned} \quad \text{or} \quad \mathbf{X} = e^{\lambda t}\begin{pmatrix} k_1 \\ k_2 \end{pmatrix} = e^{\lambda t}\mathbf{X}_1$$

Hence

$$\frac{d\mathbf{X}}{dt} = \lambda e^{\lambda t}\mathbf{X}_1 = \lambda\mathbf{X}$$

From equation (2.20) this gives

$$\frac{d\mathbf{X}}{dt} = \mathbf{AX} = \lambda\mathbf{X}$$

which is now in the form of a latent root problem. We know that in general a two by two matrix has two latent roots and two linearly independent latent vectors, so that in this case the general solution is

$$\mathbf{X} = e^{\lambda_1 t}\mathbf{X}_1 + e^{\lambda_2 t}\mathbf{X}_2$$

where λ_1 and λ_2 are the latent roots of $\mathbf{A}$, and $\mathbf{X}_1$ and $\mathbf{X}_2$ are the corresponding latent vectors.

Example 2.2

$$\frac{dx_1}{dt} = 4x_1 - 2x_2$$

$$\frac{dx_2}{dt} = 5x_1 - 3x_2$$

$$|\mathbf{A} - \lambda\mathbf{I}| = \begin{vmatrix} 4-\lambda & -2 \\ 5 & -3-\lambda \end{vmatrix} = \lambda^2 - \lambda - 2 = (\lambda - 2)(\lambda + 1) = 0$$

hence

$$\lambda_1 = 2 \quad \text{and} \quad \lambda_2 = -1$$

When $\lambda = 2$,

$$4x_1 - 2x_2 = 2x_1$$

$$5x_1 - 3x_2 = 2x_2$$

hence

$$x_1 = x_2$$

When $\lambda = -1$,

$$4x_1 - 2x_2 = -x_1$$

$$5x_1 - 3x_2 = -x_2$$

hence

$$5x_1 = 2x_2$$

so that

$$\mathbf{X}_1 = k_1\begin{pmatrix} 1 \\ 1 \end{pmatrix} \quad \text{and} \quad \mathbf{X}_2 = k_2\begin{pmatrix} 2 \\ 5 \end{pmatrix}$$

and the general solution is

$$\mathbf{X} = \begin{pmatrix} x_1 \\ x_2 \end{pmatrix} = k_1 e^{2t}\begin{pmatrix} 1 \\ 1 \end{pmatrix} + k_2 e^{-t}\begin{pmatrix} 2 \\ 5 \end{pmatrix}$$

If the matrix $\mathbf{A}$ has equal roots then it may not have n linearly independent latent vectors and the solution is not quite so straightforward. We can approach the problem by using theorem 1.5. We wish to solve the equation

$$\frac{d\mathbf{X}}{dt} = \mathbf{AX}$$

Let **C** be a unitary matrix such that $\mathbf{B} = \mathbf{C}^{-1}\mathbf{AC}$ where **B** is triangular. Then put

$$\mathbf{Y} = \mathbf{C}^{-1}\mathbf{X} \quad \text{or} \quad \mathbf{X} = \mathbf{CY}$$

so that the equation becomes

$$\frac{\mathbf{C}\,d\mathbf{Y}}{dt} = \mathbf{ACY}$$

which gives

$$\frac{d\mathbf{Y}}{dt} = \mathbf{C}^{-1}\mathbf{ACY} = \mathbf{BY}$$

Since **B** is triangular this equation may easily be solved for **Y**. An alternative approach is to use the Jordan cononical form which is not dealt with in this book.†

Example 2.3

$$\frac{dx_1}{dt} = 14x_1 - 9x_2$$

$$\frac{dx_2}{dt} = 16x_1 - 10x_2$$

$$|\mathbf{A} - \lambda\mathbf{I}| = \begin{vmatrix} 14-\lambda & -9 \\ 16 & -10-\lambda \end{vmatrix} = \lambda^2 - 4\lambda + 4 = (\lambda-2)^2 = 0$$

When $\lambda = 2$,

$$14x_1 - 9x_2 = 2x_1$$

$$16x_1 - 10x_2 = 2x_2$$

Hence

$$4x_1 = 3x_2$$

and **A** has only one linearly independent latent vector. Using theorem 1.5 we get

$$\mathbf{X}_1 = \begin{pmatrix} \frac{3}{5} \\ \frac{4}{5} \end{pmatrix} \quad (\mathbf{X}_1^T \mathbf{X}_1 = 1)$$

so that

$$\mathbf{C} = \begin{pmatrix} \frac{3}{5} & -\frac{4}{5} \\ \frac{4}{5} & \frac{3}{5} \end{pmatrix}$$

† See reference 7, Chapter 1, §§ 8, 28 and 29 or reference 19, Chapter 5.

hence

$$\mathbf{B} = \mathbf{C}^{-1}\mathbf{AC} = \begin{pmatrix} \frac{3}{5} & \frac{4}{5} \\ -\frac{4}{5} & \frac{3}{5} \end{pmatrix} \begin{pmatrix} 14 & -9 \\ 16 & -10 \end{pmatrix} \begin{pmatrix} \frac{3}{5} & -\frac{4}{5} \\ \frac{4}{5} & \frac{3}{5} \end{pmatrix}$$

$$= \begin{pmatrix} 2 & -25 \\ 0 & 2 \end{pmatrix}$$

Putting $\mathbf{X} = \mathbf{CY}$ the original equations become

$$\frac{d}{dt}\begin{pmatrix} y_1 \\ y_2 \end{pmatrix} = \begin{pmatrix} 2 & -25 \\ 0 & 2 \end{pmatrix} \begin{pmatrix} y_1 \\ y_2 \end{pmatrix}$$

The solution to the second of these is

$$y_2 = k_2 \mathrm{e}^{2t}$$

and hence the first equation becomes

$$\frac{dy_1}{dt} = 2y_1 - 25y_2 = 2y_1 - 25k_2 \mathrm{e}^{2t}$$

or

$$\frac{dy_1}{dt} - 2y_1 = -25k_2 \mathrm{e}^{2t}$$

which has the solution

$$y_1 = k_1 \mathrm{e}^{2t} - 25k_2 t\mathrm{e}^{2t}$$

Hence

$$\mathbf{Y} = \begin{pmatrix} k_1 - 25k_2 t \\ k_2 \end{pmatrix} \mathrm{e}^{2t}$$

and since $\mathbf{X} = \mathbf{CY}$,

$$\mathbf{X} = \begin{pmatrix} \frac{3}{5} & -\frac{4}{5} \\ \frac{4}{5} & \frac{3}{5} \end{pmatrix} \begin{pmatrix} k_1 - 25k_2 t \\ k_2 \end{pmatrix} \mathrm{e}^{2t} = \begin{pmatrix} \frac{1}{5}(3k_1 - 4k_2) - 15k_2 t \\ \frac{1}{5}(4k_1 + 3k_2) - 20k_2 t \end{pmatrix} \mathrm{e}^{2t}$$

Clearly these methods are easily extended to deal with more than two equations. Equations of the form

$$\frac{d^r\mathbf{X}}{dt^r} = \mathbf{AX}$$

may also be easily solved in the same way. Equations of the form

$$\frac{d^r\mathbf{X}}{dt^r} + \mathbf{B}_1\frac{d^{r-1}\mathbf{X}}{dt^{r-1}} + \ldots + \mathbf{B}_{r-1}\frac{d\mathbf{X}}{dt} = \mathbf{AX}$$

need slightly more sophisticated methods.† Note that the equation,

$$\mathbf{B}\frac{d\mathbf{X}}{dt} = \mathbf{CX}$$

† See reference 7, Chapter 1, § 30.

can be transformed into standard form by putting

$$\mathbf{A} = \mathbf{B}^{-1}\mathbf{C}$$

providing that $\mathbf{B}^{-1}$ exists. If $\mathbf{B}$ is singular the equations may be reduced to a smaller set.

2.5 Exercises

Section 2.1

2.1. Find the axes of symmetry of the hyperbola

$$31x^2+48xy+17y^2 = 1$$

Find also the distances from the origin to the intersections of these axes and the hyperbola.

2.2. Find the axes of symmetry of the ellipsoid

$$8x^2+29y^2+29z^2+28xy+28xz+56yz = 1$$

Find also the distances from the origin to the intersections of these axes and the ellipsoid.

2.3. Show that if the axes of the general conic

$$ax^2+2hxy+by^2+2fx+2gy+c = 0$$

are rotated to lie parallel with the x, y-axes, the equation takes the form

$$\lambda_1 x^2+\lambda_2 y^2+px+qy+c = 0$$

where λ_1 and λ_2 are the latent roots of the matrix

$$\begin{pmatrix} a & h \\ h & b \end{pmatrix}$$

Give conditions for λ_1 and λ_2 that determine whether the conic is an ellipse or a circle or a hyperbola or a parabola.

2.4. Determine the nature of the following conic sections

$$\text{(i)}\quad x^2+4xy-2y^2+6x-8y = 1$$

$$\text{(ii)}\quad 4x^2+12xy+9y^2-x+2y = 1$$

2.5. Extend the results of exercise 2.3 to three-dimensional conics.

Section 2.2

2.6. Solve the following equations by both the Jacobi and the Gauss–Seidel methods taking $x_1 = x_2 = x_3 = x_4 = 0$ as the initial approximation.

$$\begin{aligned} 2x_1 \qquad\quad + \; 3x_3 + \; 3x_4 &= \; 3{\cdot}0 \\ -2x_1 - \; x_2 + 0{\cdot}1x_3 + 0{\cdot}1x_4 &= -7{\cdot}0 \\ 40x_1 + 20x_2 - 200x_3 + \; 2x_4 &= \; 39{\cdot}0 \\ 20x_1 + 10x_2 - \; x_3 + 100x_4 &= \; 19{\cdot}5 \end{aligned}$$

2.7. Determine whether or not the Jacobi and Gauss–Seidel methods will converge for the following matrices of coefficients

$$\text{(i)}\quad \mathbf{A} = \begin{pmatrix} 1 & 1 & 1 \\ 2 & 1 & 1 \\ -399 & 199 & 100 \end{pmatrix} \qquad \text{(ii)}\quad \mathbf{A} = \begin{pmatrix} 1 & 0 & 1 \\ 82 & 1 & 1 \\ 80 & 1 & 10 \end{pmatrix}$$

and use the appropriate method to solve the equations,

$$\text{(iii)}\quad \begin{aligned} x+y+z &= 3 \\ 2x+y+z &= 4 \\ -399x+199y+100z &= -100 \end{aligned} \qquad \text{(iv)}\quad \begin{aligned} x+z &= 2 \\ 82x+y+z &= 84 \\ 80x+y+10z &= 91 \end{aligned}$$

2.8. If $\mathbf{X}^T\mathbf{A}\mathbf{X} > 0$ for every non-zero real vector $\mathbf{X}$ then $\mathbf{A}$ is said to be positive definite.

Prove that if $\mathbf{A}$ is a symmetric positive definite matrix then the Gauss–Seidel method converges.

(This type of matrix often occurs in practical examples. One such case is the normal matrix of coefficients obtained in the method of least squares regression.)

Section 2.3

2.9. Use the Crank–Nicolson method to obtain a numerical solution when $x = 0, 0{\cdot}1, 0{\cdot}2, 0{\cdot}3, 0{\cdot}4, 0{\cdot}5$ and $t = 0{\cdot}01$ and $0{\cdot}02$ to the equation

$$\frac{\partial u}{\partial t} = \frac{\partial^2 u}{\partial x^2} \quad (0 \leqslant x \leqslant 1)$$

with the initial condition

$$u = \cos \pi(x+\tfrac{1}{2}) \quad \text{when } t = 0$$

and the boundary conditions

$$u = 0 \quad \text{when } x = 0 \text{ or } 1 \quad \text{and } t > 0$$

Compare the numerical solution at these points with the analytical solution

$$u = e^{-\pi^2 t} \cos \pi(x+\tfrac{1}{2})$$

(There is clearly no advantage in obtaining a numerical solution in this particular example, but many partial differential equations have no known analytic solution and many others have extremely cumbersome analytic solutions.†)

2.10. Show that the finite difference approximation to

$$\frac{\partial u}{\partial t} = \frac{\partial^2 u}{\partial x^2}$$

given by equations (2.15) and (2.16) is stable when

$$k/h^2 = r < \tfrac{1}{2}$$

† See for example reference 4, p. 3.

2.11. Show that both the Jacobi and Gauss–Seidel methods will converge when applied to the equations obtained by the Crank–Nicolson method for all values of r.

Section 2.4

2.12. Solve the equations

(i) $$\frac{dx_1}{dt} = 4x_1 + x_2$$
$$\frac{dx_2}{dt} = x_1 + 4x_2$$

(ii) $$\frac{dx_1}{dt} = 3x_1 + x_2$$
$$\frac{dx_2}{dt} = -5x_1 - x_2$$

(iii) $$\frac{dx_1}{dt} = 5x_1 - x_2$$
$$\frac{dx_2}{dt} = 7x_1 - 3x_2$$

given that $x_1 = 5$ and $x_2 = 17$ when $t = 0$.

(iv) $$\frac{dx_1}{dt} = x_1 - 2x_2$$
$$\frac{dx_2}{dt} = 8x_1 - 7x_2$$

(v) $$\frac{dx_1}{dt} = x_1 - x_2$$
$$\frac{dx_2}{dt} = x_1 + 3x_2$$

given that $x_1 = x_2 = \sin \pi/4$ when $t = 0$.

(vi) $$\frac{3dx_1}{dt} + \frac{dx_2}{dt} = 5x_1 + x_2$$
$$\frac{4dx_1}{dt} + \frac{2dx_2}{dt} = 8x_1 + 4x_2$$

(vii)

$$\begin{pmatrix} 0 & -2 & -1 \\ 2 & -1 & -2 \\ -1 & 2 & 2 \end{pmatrix} \begin{pmatrix} dx_1/dt \\ dx_2/dt \\ dx_3/dt \end{pmatrix} = \begin{pmatrix} -4 & -1 & 4 \\ 1 & 1 & -1 \\ 3 & 1 & -3 \end{pmatrix} \begin{pmatrix} x_1 \\ x_2 \\ x_3 \end{pmatrix}$$

2.13. (i) Show that the characteristic equation of

$$\mathbf{A} = \begin{pmatrix} -10 & 3 & 3 \\ -50 & 17 & 10 \\ 48 & -19 & -3 \end{pmatrix}$$

is

$$\lambda^3-4\lambda^2+5\lambda-2 = 0$$

and that $\mathbf{X}_1 = k\begin{pmatrix} 1 \\ 2 \\ 2 \end{pmatrix}$ is a latent vector.

Show that

$$\mathbf{C}_1 = \tfrac{1}{3}\begin{pmatrix} 1 & 2 & 2 \\ 2 & -2 & 1 \\ 2 & 1 & -2 \end{pmatrix}$$

is an orthogonal matrix. ($\mathbf{X}_1$ is the first column and the remaining two columns are orthogonal to $\mathbf{X}_1$ and to each other.) Show also that

$$\mathbf{A}_1 = \mathbf{C}_1^{-1}\,\mathbf{A}\mathbf{C}_1 = \begin{pmatrix} 2 & -1 & -7 \\ 0 & 37 & 27 \\ 0 & -48 & -35 \end{pmatrix}$$

Put

$$\mathbf{A}_2 = \begin{pmatrix} 37 & 27 \\ -48 & -35 \end{pmatrix}$$

What are the latent roots of $\mathbf{A}_2$? Find the latent vector of $\mathbf{A}_2$. Hence find an orthogonal matrix, $\mathbf{C}_2$, such that

$$\mathbf{B} = \mathbf{C}_2^{-1}\,\mathbf{A}_1\,\mathbf{C}_2 = \begin{pmatrix} 2 & 5 & -5 \\ 0 & 1 & 75 \\ 0 & 0 & 1 \end{pmatrix}$$

(See theorem 1.5 and compare the proof with the finding of **B** here.) Notice that $\mathbf{B} = \mathbf{C}_2^{-1}\,\mathbf{C}_1^{-1}\,\mathbf{A}\mathbf{C}_1\,\mathbf{C}_2 = (\mathbf{C}_1\,\mathbf{C}_2)^{-1}\,\mathbf{A}(\mathbf{C}_1\,\mathbf{C}_2)$.

(ii) Solve the equations $d\mathbf{X}/dt = \mathbf{A}\mathbf{X}$ for the matrix **A** of (i).

(iii) Solve the equations

(a)
$$\begin{pmatrix} dx_1/dt \\ dx_2/dt \\ dx_3/dt \end{pmatrix} = \begin{pmatrix} -1 & 1 & -4 \\ 1 & 3 & 1 \\ 2 & -2 & 5 \end{pmatrix}\begin{pmatrix} x_1 \\ x_2 \\ x_3 \end{pmatrix}$$

(b)
$$\begin{pmatrix} dx_1/dt \\ dx_2/dt \\ dx_3/dt \end{pmatrix} = \begin{pmatrix} 5 & -1 & -1 \\ -4 & -7 & 10 \\ 2 & -5 & 5 \end{pmatrix}\begin{pmatrix} x_1 \\ x_2 \\ x_3 \end{pmatrix}$$

2.14. There are n particles, each of mass m, at equal distances l along a light elastic string of length $(n+1)l$ whose ends are fixed. Show that if the particles execute small transverse vibrations under no external forces then the equations of motion of the system are given by

$$\frac{T}{l}\begin{pmatrix} -2 & 1 & 0 & \dots & 0 \\ 1 & -2 & 1 & \dots & 0 \\ 0 & 1 & -2 & \dots & 0 \\ \vdots & \vdots & \vdots & & \vdots \\ 0 & 0 & 0 & \dots & -2 \end{pmatrix}\begin{pmatrix} x_1 \\ x_2 \\ x_3 \\ \vdots \\ x_n \end{pmatrix} = m\begin{pmatrix} d^2x_1/dt^2 \\ d^2x_2/dt^2 \\ d^2x_3/dt^2 \\ \vdots \\ d^2x_n/dt^2 \end{pmatrix}$$

where x_i is the displacement of the ith particle after time t and T is the tension in the string. Hence show that the frequency of vibration of the system is

$$w = \sqrt{\left(\frac{\lambda T}{ml}\right)}$$

and the mode of vibration is given by the ratios

$$y_1 : y_2 : \dots : y_n$$

where λ is a latent root, and y_i is the ith element of the corresponding latent vector of the $n \times n$ matrix

$$\mathbf{A} = \begin{pmatrix} 2 & -1 & 0 & \dots & 0 \\ -1 & 2 & -1 & \dots & 0 \\ 0 & -1 & 2 & \dots & 0 \\ \vdots & \vdots & \vdots & & \vdots \\ 0 & 0 & 0 & \dots & 2 \end{pmatrix}$$

Find the latent roots and vectors of **A**.† The theory of vibration is one of the important applications of simultaneous differential equations.‡

2.15. Show that the general solution of the simultaneous linear difference equations

$$\mathbf{X}_{r+1} = \mathbf{A}\mathbf{X}_r$$

is given by

$$\mathbf{X}_r = \lambda_1^r \mathbf{Y}_1 + \lambda_2^r \mathbf{Y}_2 + \dots + \lambda_n^r \mathbf{Y}_n$$

where λ_i is the ith latent root of **A** and $\mathbf{Y}_i$ is the corresponding latent vector and **A** has n distinct latent roots.

2.16. Solve the equations

(i)
$$x_{n+1} = x_n + 2y_n$$
$$y_{n+1} = 2x_n + y_n$$

(ii)
$$x_{n+1} = x_n - 2y_n$$
$$y_{n+1} = 2x_n - 3y_n$$

given that $x_1 = 0$ and $y_1 = 1$.

† See Appendix 1.
‡ See reference 20.

Chapter 3

THE METHOD OF DANILEVSKY

The method of Danilevsky finds the characteristic equation of a matrix by attempting to reduce it, using similarity transformations, to a Frobenius matrix which is now defined.

3.1 Frobenius Matrix

A Frobenius matrix is a matrix of the form

$$\mathbf{B} = \begin{pmatrix} b_1 & b_2 & b_3 & \ldots & b_{n-1} & b_n \\ 1 & 0 & 0 & \ldots & 0 & 0 \\ 0 & 1 & 0 & \ldots & 0 & 0 \\ \vdots & \vdots & \vdots & & \vdots & \vdots \\ 0 & 0 & 0 & \ldots & 1 & 0 \end{pmatrix}$$

B has the important property that the elements in its first row are the coefficients of its characteristic equation, because

$$|\mathbf{B}-\lambda\mathbf{I}| = \begin{vmatrix} b_1-\lambda & b_2 & b_3 & \ldots & b_{n-1} & b_n \\ 1 & -\lambda & 0 & \ldots & 0 & 0 \\ 0 & 1 & -\lambda & \ldots & 0 & 0 \\ \vdots & \vdots & \vdots & & \vdots & \vdots \\ 0 & 0 & 0 & \ldots & 1 & -\lambda \end{vmatrix} = 0$$

and expanding along the first row we get

$$\begin{aligned} |\mathbf{B}-\lambda\mathbf{I}| &= (b_1-\lambda)(-\lambda)^{n-1} - b_2(1)(-\lambda)^{n-2} + b_3(1)^2(-\lambda)^{n-3} \ldots \\ &\qquad + (-1)^n b_{n-1}(1)^{n-2}(-\lambda) + (-1)^{n+1} b_n(1)^{n-1} \\ &= (-1)^n(\lambda^n - b_1\lambda^{n-1} - b_2\lambda^{n-2} - \ldots - b_{n-1}\lambda - b_n) = 0 \end{aligned}$$

Hence the characteristic equation of **B** is

$$\lambda^n - b_1\lambda^{n-1} - b_2\lambda^{n-2} - \ldots - b_{n-1}\lambda - b_n = 0$$

B is said to be the companion matrix of any matrix to which it is similar.

3.2 Method of Danilevsky

We wish to reduce $\mathbf{A}$ given by

$$\mathbf{A} = \begin{pmatrix} a_{11} & a_{12} & \cdots & a_{1n} \\ a_{21} & a_{22} & \cdots & a_{2n} \\ \vdots & \vdots & & \vdots \\ a_{n1} & a_{n2} & \cdots & a_{nn} \end{pmatrix}$$

to its companion matrix by means of similarity transformations.

We first reduce the nth row to the required form by defining a matrix $\mathbf{C}_{n-1}$, assuming initially that $a_{n,n-1} \neq 0$:

$$\mathbf{C}_{n-1} = \begin{pmatrix} 1 & 0 & \cdots & 0 & 0 \\ 0 & 1 & \cdots & 0 & 0 \\ \vdots & \vdots & & \vdots & \vdots \\ -\dfrac{a_{n1}}{a_{n,n-1}} & -\dfrac{a_{n2}}{a_{n,n-1}} & \cdots & \dfrac{1}{a_{n,n-1}} & -\dfrac{a_{nn}}{a_{n,n-1}} \\ 0 & 0 & \cdots & 0 & 1 \end{pmatrix}$$

which means that

$$\mathbf{C}_{n-1}^{-1} = \begin{pmatrix} 1 & 0 & \cdots & 0 & 0 \\ 0 & 1 & \cdots & 0 & 0 \\ \vdots & \vdots & & \vdots & \vdots \\ a_{n1} & a_{n2} & \cdots & a_{n,n-1} & a_{nn} \\ 0 & 0 & \cdots & 0 & 1 \end{pmatrix}$$

Now, $\mathbf{AC}_{n-1}$ is a matrix of the form

$$\mathbf{AC}_{n-1} = \begin{pmatrix} b_{11} & b_{12} & \cdots & b_{1,n-1} & b_{nn} \\ b_{21} & b_{22} & \cdots & b_{2,n-1} & b_{2n} \\ \vdots & \vdots & & \vdots & \vdots \\ b_{n-1,1} & b_{n-2,2} & \cdots & b_{n-1,n-1} & b_{n-1,n} \\ 0 & 0 & \cdots & 1 & 0 \end{pmatrix}$$

where

$$b_{ij} = a_{ij} - \frac{a_{i,n-1} \cdot a_{nj}}{a_{n,n-1}} \quad \text{for all } i \leqslant n-1 \text{ and } j \neq n-1$$

and

$$b_{i,n-1} = \frac{a_{i,n-1}}{a_{n,n-1}}$$

This gives $\mathbf{C}_{n-1}^{-1}\mathbf{A}\mathbf{C}_{n-1}$ as a matrix of the form

$$\mathbf{C}_{n-1}^{-1}\mathbf{A}\mathbf{C}_{n-1} = \mathbf{A}_1 = \begin{pmatrix} b_{11} & b_{12} & \dots & b_{1,n-1} & b_{1n} \\ b_{21} & b_{22} & \dots & b_{2,n-1} & b_{2n} \\ \vdots & \vdots & & \vdots & \vdots \\ c_1 & c_2 & \dots & c_{n-1} & c_n \\ 0 & 0 & \dots & 1 & 0 \end{pmatrix}$$

where

$$c_j = \sum_{i=1}^{n-1} a_{ni} \cdot b_{ij} \quad \text{for all } j \neq n-1$$

and

$$c_{n-1} = \sum_{i=1}^{n-1} a_{ni} \cdot b_{i,n-1} + a_{nn}$$

We have now found a matrix $\mathbf{A}_1$ which is similar to $\mathbf{A}$ and has the nth row in the required form. To reduce the $(n-1)$th row of $\mathbf{A}_1$ to the required form we define a matrix $\mathbf{C}_{n-2}$, assuming $c_{n-2} \neq 0$:

$$\mathbf{C}_{n-2} = \begin{pmatrix} 1 & 0 & \dots & 0 & 0 & 0 \\ 0 & 1 & \dots & 0 & 0 & 0 \\ \vdots & \vdots & & \vdots & \vdots & \vdots \\ -\dfrac{c_1}{c_{n-2}} & -\dfrac{c_2}{c_{n-2}} & \dots & \dfrac{1}{c_{n-2}} & -\dfrac{c_{n-1}}{c_{n-2}} & -\dfrac{c_n}{c_{n-2}} \\ 0 & 0 & \dots & 0 & 1 & 0 \\ 0 & 0 & \dots & 0 & 0 & 1 \end{pmatrix}$$

which means that

$$\mathbf{C}_{n-2}^{-1} = \begin{pmatrix} 1 & 0 & \dots & 0 & 0 & 0 \\ 0 & 1 & \dots & 0 & 0 & 0 \\ \vdots & \vdots & & \vdots & \vdots & \vdots \\ c_1 & c_2 & \dots & c_{n-2} & c_{n-1} & c_n \\ 0 & 0 & \dots & 0 & 1 & 0 \\ 0 & 0 & \dots & 0 & 0 & 1 \end{pmatrix}$$

We have

$$\mathbf{A}_1\mathbf{C}_{n-2} = \begin{pmatrix} c_{11} & c_{12} & \dots & c_{1,n-2} & c_{1,n-1} & c_{1n} \\ c_{21} & c_{22} & \dots & c_{2,n-2} & c_{2,n-1} & c_{2n} \\ \vdots & \vdots & & \vdots & \vdots & \vdots \\ 0 & 0 & \dots & 1 & 0 & 0 \\ 0 & 0 & \dots & 0 & 1 & 0 \end{pmatrix}$$

where

$$c_{ij} = b_{ij} - \frac{b_{i,n-2} \cdot c_j}{c_{n-2}} \quad \text{for all } i \leqslant n-2,\ j \neq n-2$$

and

$$c_{i,n-2} = \frac{b_{i,n-2}}{c_{n-2}}$$

so that

$$\mathbf{C}_{n-2}^{-1}\,\mathbf{A}_1\,\mathbf{C}_{n-2} = \mathbf{A}_2 = \begin{pmatrix} c_{11} & c_{12} & \dots & c_{1,n-2} & c_{1,n-1} & c_{1n} \\ c_{21} & c_{22} & \dots & c_{2,n-2} & c_{2,n-1} & c_{2n} \\ \vdots & \vdots & & \vdots & \vdots & \vdots \\ d_1 & d_2 & \dots & d_{n-2} & d_{n-1} & d_n \\ 0 & 0 & \dots & 1 & 0 & 0 \\ 0 & 0 & \dots & 0 & 1 & 0 \end{pmatrix}$$

where

$$d_j = \sum_{i=1}^{n-2} c_i \cdot c_{ij} \quad \text{for all } j \neq n-2 \text{ or } n-1$$

and

$$d_j = \sum_{i=1}^{n-2} c_i \cdot c_{ij} + c_j \quad \text{for } j = n-2,\ n-1$$

It is important to notice that in forming $\mathbf{A}_2$ we have not altered the nth row of $\mathbf{A}_1$.

We proceed in this manner until we achieve the Frobenius matrix $\mathbf{B}$.

Example 3.1

$$\mathbf{A} = \begin{pmatrix} 0 & -2 & 5 \\ -7 & 1 & 9 \\ -1 & -2 & 6 \end{pmatrix}$$

Then

$$\mathbf{C}_2 = \begin{pmatrix} 1 & 0 & 0 \\ -\frac{1}{2} & -\frac{1}{2} & 3 \\ 0 & 0 & 1 \end{pmatrix} \quad \text{and} \quad \mathbf{C}_2^{-1} = \begin{pmatrix} 1 & 0 & 0 \\ -1 & -2 & 6 \\ 0 & 0 & 1 \end{pmatrix}$$

Hence

$$\mathbf{AC}_2 = \begin{pmatrix} 0 & -2 & 5 \\ -7 & 1 & 9 \\ -1 & -2 & 6 \end{pmatrix} \begin{pmatrix} 1 & 0 & 0 \\ -\frac{1}{2} & -\frac{1}{2} & 3 \\ 0 & 0 & 1 \end{pmatrix} = \begin{pmatrix} 1 & 1 & -1 \\ -\frac{15}{2} & -\frac{1}{2} & 12 \\ 0 & 1 & 0 \end{pmatrix}$$

and

$$\mathbf{C}_2^{-1}\mathbf{A}\mathbf{C}_2 = \begin{pmatrix} 1 & 0 & 0 \\ -1 & -2 & 6 \\ 0 & 0 & 1 \end{pmatrix} \begin{pmatrix} 1 & 1 & -1 \\ -\frac{15}{2} & -\frac{1}{2} & 12 \\ 0 & 1 & 0 \end{pmatrix} = \begin{pmatrix} 1 & 1 & -1 \\ 14 & 6 & -23 \\ 0 & 1 & 0 \end{pmatrix} = \mathbf{A}_1$$

so

$$\mathbf{C}_1 = \begin{pmatrix} \frac{1}{14} & -\frac{3}{7} & \frac{23}{14} \\ 0 & 1 & 0 \\ 0 & 0 & 1 \end{pmatrix} \quad \text{and} \quad \mathbf{C}_1^{-1} = \begin{pmatrix} 14 & 6 & -23 \\ 0 & 1 & 0 \\ 0 & 0 & 1 \end{pmatrix}$$

which gives

$$\mathbf{A}_1\mathbf{C}_1 = \begin{pmatrix} 1 & 1 & -1 \\ 14 & 6 & -23 \\ 0 & 1 & 0 \end{pmatrix} \begin{pmatrix} \frac{1}{14} & -\frac{3}{7} & \frac{23}{14} \\ 0 & 1 & 0 \\ 0 & 0 & 1 \end{pmatrix} = \begin{pmatrix} \frac{1}{14} & \frac{4}{7} & \frac{9}{14} \\ 1 & 0 & 0 \\ 0 & 1 & 0 \end{pmatrix}$$

and

$$\mathbf{C}_1^{-1}\mathbf{A}_1\mathbf{C}_1 = \begin{pmatrix} 14 & 6 & -23 \\ 0 & 1 & 0 \\ 0 & 0 & 1 \end{pmatrix} \begin{pmatrix} \frac{1}{14} & \frac{4}{7} & \frac{9}{14} \\ 1 & 0 & 0 \\ 0 & 1 & 0 \end{pmatrix} = \begin{pmatrix} 7 & -15 & 9 \\ 1 & 0 & 0 \\ 0 & 1 & 0 \end{pmatrix} = \mathbf{A}_2 = \mathbf{B}$$

Therefore the characteristic equation of **A** is

$$\lambda^3 - 7\lambda^2 + 15\lambda - 9 = 0$$

which has roots

$$\lambda_1 = 1, \quad \lambda_2 = \lambda_3 = 3$$

3.3 Pivot Element Equal to Zero—Case 1

In the above section we made the assumption that the element we wished to divide by was non-zero. This, of course, is not in general true.

Suppose that in the Danilevsky process we reach the matrix $\mathbf{A}_{n-r}$ given by

$$\mathbf{A}_{n-r} = \begin{pmatrix} a_{11} & \cdots & a_{1j} & \cdots & a_{1,r-1} & a_{1r} & \cdots & a_{1,n-1} & a_{1n} \\ \vdots & & \vdots & & \vdots & \vdots & & \vdots & \vdots \\ a_{j1} & \cdots & a_{jj} & \cdots & a_{j,r-1} & a_{jr} & \cdots & a_{j,n-1} & a_{jn} \\ \vdots & & \vdots & & \vdots & \vdots & & \vdots & \vdots \\ a_{r-1,1} & \cdots & a_{r-1,j} & \cdots & a_{r-1,r-1} & a_{r-1,r} & \cdots & a_{r-1,n-1} & a_{r-1,n} \\ a_{r1} & \cdots & a_{rj} & \cdots & 0 & a_{rr} & \cdots & a_{r,n-1} & a_{rn} \\ 0 & \cdots & 0 & \cdots & 0 & 1 & \cdots & 0 & 0 \\ \vdots & & \vdots & & \vdots & \vdots & & \vdots & \vdots \\ 0 & \cdots & 0 & \cdots & 0 & 0 & \cdots & 1 & 0 \end{pmatrix}$$

Here the element we wish to divide by in forming $\mathbf{C}_{r-1}$ is zero. Suppose that $a_{rj} \neq 0$ for some $j < r-1$, then we define a similarity transformation that will interchange $a_{r,r-1}$ and a_{rj}.

In order to do this we define a matrix $\mathbf{S}$ that will interchange the $(r-1)$th and jth columns of $\mathbf{A}_{n-r}$ when we form $\mathbf{A}_{n-r}\mathbf{S}$. This means that $\mathbf{S}$ is given by

$$\mathbf{S} = \begin{pmatrix} 1 & 0 & \dots & 0 & \dots & 0 & \dots & 0 & 0 \\ 0 & 1 & \dots & 0 & \dots & 0 & \dots & 0 & 0 \\ \vdots & \vdots & \ddots & \vdots & & \vdots & ⋰ & \vdots & \vdots \\ 0 & 0 & \dots & 0 & \dots & 1 & \dots & 0 & 0 \\ \vdots & \vdots & & \vdots & ∵ & \vdots & & \vdots & \vdots \\ 0 & 0 & \dots & 1 & \dots & 0 & \dots & 0 & 0 \\ \vdots & \vdots & ⋰ & \vdots & & \vdots & \ddots & \vdots & \vdots \\ 0 & 0 & \dots & 0 & \dots & 0 & \dots & 1 & 0 \\ 0 & 0 & \dots & 0 & \dots & 0 & \dots & 0 & 1 \end{pmatrix} \begin{matrix} \\ \\ \\ j\text{th row} \\ \\ (r-1)\text{th row} \\ \\ \\ \\ \end{matrix}$$

jth column, $(r-1)$th column

It is easy to see that $\mathbf{S}^{-1} = \mathbf{S}$ and hence $\mathbf{S}^{-1}\mathbf{A}_{n-r}\mathbf{S} = \mathbf{S}\mathbf{A}_{n-r}\mathbf{S}$ will have the effect of interchanging the jth and $(r-1)$th rows of $\mathbf{A}_{n-r}\mathbf{S}$. This, of course, means that given $\mathbf{A}_{n-r}$ we can write down $\mathbf{S}^{-1}\mathbf{A}_{n-r}\mathbf{S}$ without actually having to perform the matrix multiplications. Furthermore, it is clear that this process does not alter the $(\mathbf{r}+1)$th, $(\mathbf{r}+2)$th, ..., nth rows of $\mathbf{A}_{n-r}$ as we naturally require. For this reason we cannot choose $j > r-1$.

Example 3.2

$$\mathbf{A} = \begin{pmatrix} -2 & 2 & 3 \\ -3 & 3 & 3 \\ -2 & 0 & 5 \end{pmatrix}$$

We cannot directly form $\mathbf{C}_2$ because $a_{32} = 0$, but since $a_{31} \neq 0$ we can interchange these two elements. So

$$\mathbf{AS} = \begin{pmatrix} -2 & 2 & 3 \\ -3 & 3 & 3 \\ -2 & 0 & 5 \end{pmatrix} \begin{pmatrix} 0 & 1 & 0 \\ 1 & 0 & 0 \\ 0 & 0 & 1 \end{pmatrix} = \begin{pmatrix} 2 & -2 & 3 \\ 3 & -3 & 3 \\ 0 & -2 & 5 \end{pmatrix}$$

and

$$\mathbf{S}^{-1}\mathbf{AS} = \begin{pmatrix} 0 & 1 & 0 \\ 1 & 0 & 0 \\ 0 & 0 & 1 \end{pmatrix} \begin{pmatrix} 2 & -2 & 3 \\ 3 & -3 & 3 \\ 0 & -2 & 5 \end{pmatrix} = \begin{pmatrix} 3 & -3 & 3 \\ 2 & -2 & 3 \\ 0 & -2 & 5 \end{pmatrix}$$

We can see that, as mentioned above, we could quite easily have written this new matrix straight down. Now

$$\mathbf{C}_2 = \begin{pmatrix} 1 & 0 & 0 \\ 0 & -\frac{1}{2} & \frac{5}{2} \\ 0 & 0 & 1 \end{pmatrix} \quad \text{and} \quad \mathbf{C}_2^{-1} = \begin{pmatrix} 1 & 0 & 0 \\ 0 & -2 & 5 \\ 0 & 0 & 1 \end{pmatrix}$$

which gives

$$\mathbf{A}_1 = \mathbf{C}_2^{-1}\mathbf{AC}_2 = \begin{pmatrix} 3 & \frac{3}{2} & -\frac{9}{2} \\ -4 & 3 & 4 \\ 0 & 1 & 0 \end{pmatrix}$$

Hence

$$\mathbf{C}_1 = \begin{pmatrix} -\frac{1}{4} & \frac{3}{4} & 1 \\ 0 & 1 & 0 \\ 0 & 0 & 1 \end{pmatrix} \quad \text{and} \quad \mathbf{C}_1^{-1} = \begin{pmatrix} -4 & 3 & 4 \\ 0 & 1 & 0 \\ 0 & 0 & 1 \end{pmatrix}$$

which gives

$$\mathbf{B} = \mathbf{C}_1\mathbf{A}_1\mathbf{C}_1^{-1} = \begin{pmatrix} 6 & -11 & 6 \\ 1 & 0 & 0 \\ 0 & 1 & 0 \end{pmatrix}$$

Therefore the characteristic equation of **A** is

$$\lambda^3 - 6\lambda^2 + 11\lambda - 6 = 0$$

which has roots

$$\lambda_1 = 1, \quad \lambda_2 = 2, \quad \lambda_3 = 3$$

3.4 Pivot Element Equal to Zero—Case 2

If in the matrix $\mathbf{A}_{n-r}$ of the previous section there does not exist an element $a_{rj} \neq 0$ for some $j < r-1$ then we cannot effect an interchange, but in this case we see that A_{n-r} is a matrix of the form

$$A_{n-r} = \left(\begin{array}{c|c} \mathbf{D}_1 & \mathbf{D}_2 \\ \hline 0 & \mathbf{D}_3 \end{array}\right)$$

where $\mathbf{D}_3$ is in Frobenius form. Then from theorem 1.14 we have

$$|\mathbf{A}_{n-r}-\lambda\mathbf{I}| = |\mathbf{A}-\lambda\mathbf{I}| = |\mathbf{D}_1-\lambda\mathbf{I}|\,|\mathbf{D}_3-\lambda\mathbf{I}|$$

Since $\mathbf{D}_3$ is in Frobenius form we can write its characteristic equation straight down. To find the characteristic equation of $\mathbf{D}_1$ we use Danilevsky's method to reduce it to Frobenius form.

Example 3.3

$$\mathbf{A} = \begin{pmatrix} 1 & -1 & 3 & 4 \\ 4 & 1 & 2 & 1 \\ 4 & 2 & 1 & -1 \\ 0 & -1 & 1 & 0 \end{pmatrix}$$

Then

$$\mathbf{C}_3 = \begin{pmatrix} 1 & 0 & 0 & 0 \\ 0 & 1 & 0 & 0 \\ 0 & 1 & 1 & 0 \\ 0 & 0 & 0 & 1 \end{pmatrix} \quad \text{and} \quad \mathbf{C}_3^{-1} = \begin{pmatrix} 1 & 0 & 0 & 0 \\ 0 & 1 & 0 & 0 \\ 0 & -1 & 1 & 0 \\ 0 & 0 & 0 & 1 \end{pmatrix}$$

which gives

$$\mathbf{C}_3^{-1}\mathbf{A}\mathbf{C}_3 = \begin{pmatrix} 1 & 2 & 3 & 4 \\ 4 & 3 & 2 & 1 \\ 0 & 0 & -1 & -2 \\ 0 & 0 & 1 & 0 \end{pmatrix} = \mathbf{A}_1$$

so that

$$\mathbf{D}_3 = \begin{pmatrix} -1 & -2 \\ 1 & 0 \end{pmatrix}$$

which has as its characteristic equation $\lambda^2+\lambda+2 = 0$, and

$$\mathbf{D}_1 = \begin{pmatrix} 1 & 2 \\ 4 & 3 \end{pmatrix}$$

We put

$$\mathbf{C}_1 = \begin{pmatrix} \frac{1}{4} & -\frac{3}{4} \\ 0 & 1 \end{pmatrix} \quad \text{and} \quad \mathbf{C}_1^{-1} = \begin{pmatrix} 4 & 3 \\ 0 & 1 \end{pmatrix}$$

giving

$$\mathbf{C}_1^{-1}\mathbf{D}_1\mathbf{C}_1 = \begin{pmatrix} 4 & 5 \\ 1 & 0 \end{pmatrix}$$

so that the characteristic equation of $\mathbf{D}_1$ is $\lambda^2 - 4\lambda - 5 = 0$. Hence the characteristic equation of $\mathbf{A}$ is given by

$$(\lambda^2 + \lambda + 2)(\lambda^2 - 4\lambda - 5) = 0$$

which has roots

$$\lambda_1 = -1, \quad \lambda_2 = 5, \quad \lambda_3 = \tfrac{1}{2}(-1 + i\sqrt{7}), \quad \lambda_4 = \tfrac{1}{2}(-1 - i\sqrt{7})$$

We can see that this condition considerably reduces the work involved.

3.5 Latent Vectors and Danilevsky's Method

If $\mathbf{Y}$ is a latent vector of the Frobenius matrix, we have

$$\mathbf{BY} = \begin{pmatrix} b_1 & b_2 & b_3 & \dots & b_{n-1} & b_n \\ 1 & 0 & 0 & \dots & 0 & 0 \\ 0 & 1 & 0 & \dots & 0 & 0 \\ \vdots & \vdots & \vdots & & \vdots & \vdots \\ 0 & 0 & 0 & \dots & 1 & 0 \end{pmatrix} \begin{pmatrix} y_1 \\ y_2 \\ y_3 \\ \vdots \\ y_n \end{pmatrix} = \lambda \begin{pmatrix} y_1 \\ y_2 \\ y_3 \\ \vdots \\ y_n \end{pmatrix} = \lambda \mathbf{Y}$$

which yields the set of equations

$$\begin{aligned} (b_1 - \lambda) y_1 + b_2 y_2 + b_3 y_3 + \dots + b_n y_n &= 0 \\ y_1 - \lambda y_2 \qquad\qquad &= 0 \\ y_2 - \lambda y_3 \qquad &= 0 \\ \ddots \qquad\qquad &\;\;\vdots \\ y_{n-1} - \lambda y_n &= 0 \end{aligned}$$

The last $(n-1)$ of these equations gives

$$\begin{aligned} y_{n-1} &= \lambda y_n \\ y_{n-2} &= \lambda y_{n-1} = \lambda^2 y_n \\ &\;\;\vdots \\ y_1 &= \lambda y_2 = \lambda^{n-1} y_n \end{aligned}$$

and substituting in the first equation in these gives

$$(b_1 - \lambda)\lambda^{n-1} y_n + b_2 \lambda^{n-2} y_n + b_3 \lambda^{n-3} y_n + \dots + b_n y_n = 0$$

or

$$(\lambda^n - b_1 \lambda^{n-1} - b_2 \lambda^{n-2} - b_3 \lambda^{n-3} - \dots - b_n) y_n = 0$$

which holds for any value of y_n. If we choose $y_n = 1$ we conveniently get

$$\begin{aligned} y_n &= 1 \\ y_{n-1} &= \lambda \\ y_{n-2} &= \lambda^2 \\ &\vdots \\ y_1 &= \lambda^{n-1} \end{aligned}$$

Using theorem 1.2 we also have

$$\mathbf{X} = \mathbf{C}_{n-1}\mathbf{C}_{n-2}\ldots\mathbf{C}_2\mathbf{C}_1\mathbf{Y}$$

which enables us to find the latent vectors of **A**.

Example 3.4

Take the matrix **A** of example 3.1 which had latent roots $\lambda_1 = 1, \lambda_2 = \lambda_3 = 3$. When $\lambda = 1$ we take **Y** as

$$\mathbf{Y} = \begin{pmatrix} \lambda^2 \\ \lambda \\ 1 \end{pmatrix} = \begin{pmatrix} 1 \\ 1 \\ 1 \end{pmatrix}$$

Hence

$$\mathbf{C}_2\mathbf{C}_1\mathbf{Y} = \begin{pmatrix} 1 & 0 & 0 \\ -\frac{1}{2} & -\frac{1}{2} & 3 \\ 0 & 0 & 1 \end{pmatrix} \begin{pmatrix} \frac{1}{14} & -\frac{3}{7} & \frac{23}{14} \\ 0 & 1 & 0 \\ 0 & 0 & 1 \end{pmatrix} \begin{pmatrix} 1 \\ 1 \\ 1 \end{pmatrix} = \begin{pmatrix} \frac{9}{7} \\ \frac{13}{7} \\ 1 \end{pmatrix} = \mathbf{X}$$

When $\lambda = 3$ we get

$$\mathbf{C}_2\mathbf{C}_1\mathbf{Y} = \begin{pmatrix} 1 & 0 & 0 \\ -\frac{1}{2} & -\frac{1}{2} & 3 \\ 0 & 0 & 1 \end{pmatrix} \begin{pmatrix} \frac{1}{14} & -\frac{3}{7} & \frac{23}{14} \\ 0 & 1 & 0 \\ 0 & 0 & 1 \end{pmatrix} \begin{pmatrix} 9 \\ 3 \\ 1 \end{pmatrix} = \begin{pmatrix} 1 \\ 1 \\ 1 \end{pmatrix} = \mathbf{X}$$

Note that **A** has only two linearly independent latent vectors.

If we have had to interchange elements in reducing **A** to Frobenius form (see § 3.3) we must, of course, take this into account in the above process.

The case of § 3.4 is not quite so simple. If the latent vectors are required, having reached the stage of

$$\mathbf{A}_{n-r} = \left(\begin{array}{c|c} \mathbf{D}_1 & \mathbf{D}_2 \\ \hline 0 & \mathbf{D}_3 \end{array}\right)$$

instead of just reducing $\mathbf{D}_1$ to Frobenius form, it is perhaps worth extending this reduction over the whole of $\mathbf{A}_{n-r}$ so that we finish up with a matrix of the form

$$\mathbf{B} = \begin{pmatrix} b_1 & b_2 & b_3 & \dots & b_{r-1} & b_r & c_{r+1} & c_{r+2} & \dots & c_{n-1} & c_n \\ 1 & 0 & 0 & \dots & 0 & 0 & 0 & 0 & \dots & 0 & 0 \\ 0 & 1 & 0 & \dots & 0 & 0 & 0 & 0 & \dots & 0 & 0 \\ \vdots & \vdots & \vdots & & \vdots & \vdots & \vdots & \vdots & & \vdots & \vdots \\ 0 & 0 & 0 & \dots & 1 & 0 & 0 & 0 & \dots & 0 & 0 \\ 0 & 0 & 0 & \dots & 0 & 0 & b_{r+1} & b_{r+2} & \dots & b_{n-1} & b_n \\ 0 & 0 & 0 & \dots & 0 & 0 & 1 & 0 & \dots & 0 & 0 \\ \vdots & \vdots & \vdots & & \vdots & \vdots & \vdots & \vdots & & \vdots & \vdots \\ 0 & 0 & 0 & \dots & 0 & 0 & 0 & 0 & \dots & 1 & 0 \end{pmatrix}$$

from which it is fairly easy to determine the latent vectors of $\mathbf{B}$.

Example 3.5

We take the matrix $\mathbf{A}_1$ of example 3.3 given by

$$\mathbf{A}_1 = \begin{pmatrix} 1 & 2 & 3 & 4 \\ 4 & 3 & 2 & 1 \\ 0 & 0 & -1 & -2 \\ 0 & 0 & 1 & 0 \end{pmatrix}$$

but we now take

$$\mathbf{C}_1 = \begin{pmatrix} \frac{1}{4} & -\frac{3}{4} & -\frac{1}{2} & -\frac{1}{4} \\ 0 & 1 & 0 & 0 \\ 0 & 0 & 1 & 0 \\ 0 & 0 & 0 & 1 \end{pmatrix} \quad \text{and} \quad \mathbf{C}_1^{-1} = \begin{pmatrix} 4 & 3 & 2 & 1 \\ 0 & 1 & 0 & 0 \\ 0 & 0 & 1 & 0 \\ 0 & 0 & 0 & 1 \end{pmatrix}$$

which gives

$$\mathbf{C}_1^{-1}\mathbf{A}_1\mathbf{C}_1 = \begin{pmatrix} 4 & 5 & 9 & 11 \\ 1 & 0 & 0 & 0 \\ 0 & 0 & -1 & -2 \\ 0 & 0 & 1 & 0 \end{pmatrix} = \mathbf{B}$$

so that

$$\mathbf{BY} = \begin{pmatrix} 4 & 5 & 9 & 11 \\ 1 & 0 & 0 & 0 \\ 0 & 0 & -1 & -2 \\ 0 & 0 & 1 & 0 \end{pmatrix} \begin{pmatrix} y_1 \\ y_2 \\ y_3 \\ y_4 \end{pmatrix} = \lambda \begin{pmatrix} y_1 \\ y_2 \\ y_3 \\ y_4 \end{pmatrix} = \lambda \mathbf{Y}$$

which gives the equations

$$\begin{aligned} 4y_1 + 5y_2 + 9y_3 + 11y_4 &= \lambda y_1 \\ y_1 \qquad\qquad\qquad &= \lambda y_2 \\ -y_3 - 2y_4 &= \lambda y_3 \\ y_3 \qquad &= \lambda y_4 \end{aligned}$$

The last two equations allow us to determine y_3 and y_4, which then allows us to find y_1 and y_2 from the first two equations.

We have

$$y_3 = \lambda y_4$$

so that

$$-\lambda y_4 - 2y_4 = \lambda^2 y_4$$

or

$$(\lambda^2 + \lambda + 2)\, y_4 = 0$$

which means that, if λ is a latent root of $\mathbf{D}_3$, y_4 is arbitrary, otherwise it (and y_3) must be zero.

We also have

$$y_1 = \lambda y_2$$

so that

$$4\lambda y_2 + 5y_2 + 9y_3 + 11y_4 = \lambda^2 y_2$$

or

$$(\lambda^2 - 4\lambda - 5)\, y_2 = 9y_3 + 11y_4$$

which means that if λ is a latent root of $\mathbf{D}_1$, y_2 is arbitrary, otherwise it is determined by the above equation.

Taking $\lambda_1 = -1$ we get

$$y_3 = y_4 = 0$$

and taking $y_2 = 1$ we get $y_1 = -1$ so that

$$\mathbf{Y}_1 = \begin{pmatrix} -1 \\ 1 \\ 0 \\ 0 \end{pmatrix}$$

Hence

$$\mathbf{C}_3\mathbf{C}_1\mathbf{Y}_1 = \begin{pmatrix} 1 & 0 & 0 & 0 \\ 0 & 1 & 0 & 0 \\ 0 & 1 & 1 & 0 \\ 0 & 0 & 0 & 1 \end{pmatrix} \begin{pmatrix} \frac{1}{4} & -\frac{3}{4} & -\frac{1}{2} & -\frac{1}{4} \\ 0 & 1 & 0 & 0 \\ 0 & 0 & 1 & 0 \\ 0 & 0 & 0 & 1 \end{pmatrix} \begin{pmatrix} -1 \\ 1 \\ 0 \\ 0 \end{pmatrix} = \begin{pmatrix} -1 \\ 1 \\ 1 \\ 0 \end{pmatrix} = \mathbf{X}_1$$

Taking $\lambda_2 = 5$ we get

$$y_3 = y_4 = 0$$

and taking $y_2 = 1$ this gives $y_1 = 5$ so that

$$\mathbf{C}_3\mathbf{C}_1\mathbf{Y}_2 = \begin{pmatrix} 1 & 0 & 0 & 0 \\ 0 & 1 & 0 & 0 \\ 0 & 1 & 1 & 0 \\ 0 & 0 & 0 & 1 \end{pmatrix} \begin{pmatrix} \frac{1}{4} & -\frac{3}{4} & -\frac{1}{2} & -\frac{1}{4} \\ 0 & 1 & 0 & 0 \\ 0 & 0 & 1 & 0 \\ 0 & 0 & 0 & 1 \end{pmatrix} \begin{pmatrix} 5 \\ 1 \\ 0 \\ 0 \end{pmatrix} = \begin{pmatrix} \frac{1}{2} \\ 1 \\ 1 \\ 0 \end{pmatrix} = \mathbf{X}_2$$

Taking $\lambda_3 = \frac{1}{2}(-1+i\sqrt{7})$ and putting $y_4 = 1$ we get

$$y_3 = \tfrac{1}{2}(-1+i\sqrt{7})$$

so that

$$\{\tfrac{1}{4}(-1+i\sqrt{7})^2 - 2(-1+i\sqrt{7}) - 5\}y_2 = \tfrac{9}{2}(-1+i\sqrt{7})+11$$

or

$$y_2 = \frac{-27-i\sqrt{7}}{16}$$

and

$$y_1 = \frac{(-1+i\sqrt{7})}{2}\cdot\frac{(-27-i\sqrt{7})}{16} = \frac{17-13i\sqrt{7}}{16}$$

so that

$$\mathbf{C}_3\mathbf{C}_1\mathbf{Y}_3 = \begin{pmatrix} 1 & 0 & 0 & 0 \\ 0 & 1 & 0 & 0 \\ 0 & 1 & 1 & 0 \\ 0 & 0 & 0 & 1 \end{pmatrix} \begin{pmatrix} \frac{1}{4} & -\frac{3}{4} & -\frac{1}{2} & -\frac{1}{4} \\ 0 & 1 & 0 & 0 \\ 0 & 0 & 1 & 0 \\ 0 & 0 & 0 & 1 \end{pmatrix} \begin{pmatrix} \frac{17-13i\sqrt{7}}{16} \\ \frac{-27-i\sqrt{7}}{16} \\ \frac{-1+i\sqrt{7}}{2} \\ 1 \end{pmatrix}$$

$$= \begin{pmatrix} \frac{49-13i\sqrt{7}}{32} \\ \frac{-27-i\sqrt{7}}{16} \\ \frac{-35+7i\sqrt{7}}{16} \\ 1 \end{pmatrix} = \mathbf{X}_3$$

Using theorem 1.19 we get

$$\mathbf{X}_4 = \begin{pmatrix} \dfrac{49+13i\sqrt{7}}{32} \\ \dfrac{-27+i\sqrt{7}}{16} \\ \dfrac{-35-7i\sqrt{7}}{16} \\ 1 \end{pmatrix}$$

and we have now found all the latent vectors of **A**.

It is hoped that this example is sufficient to demonstrate how we may proceed in general.

3.6 Improving the Accuracy of Danilevsky's Method

At each stage of the Danilevsky process we are dividing all the elements in the pivot row by the pivot element. If the pivot element is small this can obviously lead to bad rounding errors and inaccurate division. To try to avoid this it is advisable to select the largest element as the pivot. We can do this using the method described in § 3.3.

Our choice of pivot is nevertheless limited. For example, in the matrix $\mathbf{A}_{n-r}$ of § 3.3 we saw that we could only select as a pivot an element a_{rj} for some $j<r-1$. This means, of course, that division of the elements a_{rj} with $j>r-1$ could still lead to inaccuracies, and hence, as is pointed out by Wilkinson,† the second half of the Danilevsky method is basically unstable. (That is the elements arising from these possible inaccuracies.)

Fadeeva,‡ suggests that a comparison of b_1 with the trace of the original matrix **A** (see theorem 1.11) is made as a guide to the accuracy. This does not seem to be very useful since it is the element least likely to reflect the inaccuracies. A much better guide would be to compare b_n with $|\mathbf{A}|$ (see theorem 1.12), providing, of course, that this can be done accurately.

We note that selection of the largest pivot does not involve any additional computation.

3.7 Number of Calculations Required by Danilevsky's Method

To form the matrix $\mathbf{C}_{n-1}$ requires n divisions. Then to get $\mathbf{AC}_{n-1}$ requires $n(n-1)$ multiplications. $\mathbf{C}_{n-1}^{-1}$ can be written straight down without any

† See reference 7, p. 409.
‡ See reference 6, p. 173; also reference 8, Vol. 2, p. 212.

computation and to find $\mathbf{C}_{n-1}^{-1}\mathbf{A}\mathbf{C}_{n-1}$ requires $n(n-1)$ multiplications. So, in all, to obtain $\mathbf{A}_1$ requires

$$n+n(n-1)+n(n-1) = n(2n-1) \text{ calculations}$$

Similarly to find $\mathbf{A}_2$ requires

$$n+n(n-2)+n(n-2) = n(2n-3) \text{ calculations}$$

Continuing this process we find that the number of calculations required to obtain the characteristic equation is given by

$$\begin{aligned} S_n &= n(2n-1)+n(2n-3)+n(2n-5)+\ldots+5n+3n \\ &= n\{3+5+\ldots+(2n-5)+(2n-3)+(2n-1)\} \\ &= n(n-1)(n+1) = n(n^2-1) \end{aligned}$$

See Table 3.1.

TABLE 3.1

n	$n(n^2-1)$
3	24
4	60
5	120
6	210
7	336
8	504
9	720
10	990
20	7 980
50	124 950
100	999 900

TABLE 3.2

n	n^2-2
3	7
4	14
5	23
6	34
7	47
8	62
9	79
10	98
20	398
50	2498
100	9998

Obviously we cannot give the number of calculations required for solving the characteristic equation since this will depend on such factors as the method chosen, the number of iterations needed, whether or not complex or multiple roots are present, and the condition of the polynomial.

To find a latent vector of $\mathbf{B}$ requires $n-2$ multiplications (powers of λ). To calculate the latent vector of $\mathbf{A}$ from this vector requires $n(n-1)$ multiplications. So, in all, each latent vector of $\mathbf{A}$ requires

$$(n-2)+n(n-1) = n^2-2 \text{ calculations} \quad \text{(see Table 3.2)}$$

The calculations required in the modified methods are of the same order as those given above.

3.8 FURTHER COMMENTS ON DANILEVSKY'S METHOD

Danilevsky's method is an excellent method of finding the characteristic equation by hand computation providing care is taken not to lose significant accuracy, especially when dealing with small pivot elements.† The instability of the method reduces its effectiveness as a computer method.

The number of calculations required by the method compares favourably with other methods, especially when the generality of the method is considered.

3.9 Exercises

3.1. Use the method of Danilevsky to find the latent roots and vectors of the following matrices:

$$\text{(i)}\quad \mathbf{A} = \begin{pmatrix} 5 & -4 & 9 \\ 6 & -7 & 21 \\ 2 & -1 & 3 \end{pmatrix} \qquad \text{(ii)}\quad \mathbf{A} = \begin{pmatrix} 4 & -1 & 14 \\ -1 & 2 & -6 \\ -1 & 0 & -3 \end{pmatrix}$$

$$\text{(iii)}\quad \mathbf{A} = \begin{pmatrix} 3 & -3 & 11 & -1 & 1 \\ 0 & 4 & -3 & 1 & -1 \\ -3 & -4 & -9 & -2 & 6 \\ 8 & 11 & 28 & 7 & -17 \\ -1 & -2 & -3 & -1 & 2 \end{pmatrix}$$

3.2. Using Danilevsky's method with exact arithmetic shows that the characteristic equation of the matrix

$$\mathbf{A} = \begin{pmatrix} 99 & 700 & -70 \\ -14 & -99 & 10 \\ 1 & 7 & 6 \end{pmatrix}$$

is

$$\lambda^3 - 6\lambda^2 - \lambda + 6 = (\lambda - 6)(\lambda - 1)(\lambda + 1) = 0$$

Repeat the calculations using arithmetic that is correct to four significant figures (i.e. make each individual multiplication or addition correct to four significant figures) to show that this yields the equation

$$\lambda^3 - 5{\cdot}95\lambda^2 - 5{\cdot}95\lambda + 39{\cdot}5 = 0$$

and find the roots of this equation correct to two decimal places. (*Note*: The matrix multiplication should be performed in the same order as suggested in the text to yield this result.)

3.3. Show how the latent vectors of $\mathbf{A}^T$ may be found after Danilevsky's method has been applied. Find the latent vectors of $\mathbf{A}^T$ for exercise 3.1 (ii).

† See exercise 3.2.

Chapter 4

THE METHOD OF KRYLOV

The method of Krylov constructs a set of simultaneous linear equations, the solution of which gives the coefficients of a polynomial which will be either the characteristic equation or a factor of the characteristic equation.

4.1 The Method of Krylov

In the method of Krylov we take an arbitrary initial column vector $\mathbf{Y}_0 \neq 0$, and construct a sequence of column vectors using the recurrence relation

$$\mathbf{Y}_{i+1} = \mathbf{A}\mathbf{Y}_i \tag{4.1}$$

This gives

$$\begin{aligned} \mathbf{Y}_1 &= \mathbf{A}\mathbf{Y}_0 \\ \mathbf{Y}_2 &= \mathbf{A}\mathbf{Y}_1 = \mathbf{A}^2\mathbf{Y}_0 \\ &\vdots \\ \mathbf{Y}_i &= \mathbf{A}\mathbf{Y}_{i-1} = \mathbf{A}^i\mathbf{Y}_0 \end{aligned}$$

Suppose that the first r vectors of this sequence are linearly independent but that $\mathbf{Y}_r$ is linearly dependent on the preceding vectors. This must of course be true for some $r \leqslant n$. Hence we can express $\mathbf{Y}_r$ as

$$\mathbf{Y}_r = a_1\mathbf{Y}_{r-1} + a_2\mathbf{Y}_{r-2} + \ldots + a_{r-1}\mathbf{Y}_1 + a_r\mathbf{Y}_0 \tag{4.2}$$

This defines n simultaneous linear equations in the r unknowns $a_1, a_2, \ldots, a_r$ and we can select the first r equations to solve for these. The relevance of this will become apparent shortly.

Theorem 4.1

If $\mathbf{Y}_r$ is linearly dependent on the vectors $\mathbf{Y}_0, \mathbf{Y}_1, \ldots, \mathbf{Y}_{r-1}$, then all successive vectors in the sequence will also be linearly dependent on these vectors.

Proof

$$\mathbf{Y}_{r+1} = \mathbf{A}\mathbf{Y}_r$$

which, from equation (4.2), gives

$$\begin{aligned}\mathbf{Y}_{r+1} &= a_1\mathbf{A}\mathbf{Y}_{r-1}+a_2\mathbf{A}\mathbf{Y}_{r-2}+\ldots+a_{r-1}\mathbf{A}\mathbf{Y}_1+a_r\mathbf{A}\mathbf{Y}_0\\ &= a_1\mathbf{Y}_r+a_2\mathbf{Y}_{r-1}+\ldots+a_{r-1}\mathbf{Y}_2+a_r\mathbf{Y}_1\\ &= a_1(a_1\mathbf{Y}_{r-1}+a_2\mathbf{Y}_{r-2}+\ldots+a_{r-1}\mathbf{Y}_1+a_r\mathbf{Y}_0)+a_2\mathbf{Y}_{r-1}+\ldots+a_{r-1}\mathbf{Y}_2+a_r\mathbf{Y}_1\end{aligned}$$

so that $\mathbf{Y}_{r+1}$ is linearly dependent upon $\mathbf{Y}_0, \mathbf{Y}_1, \ldots, \mathbf{Y}_{r-1}$. Now assume that the theorem is true for all vectors in the sequence up to $\mathbf{Y}_p$, where $p > r$. Hence we can express $\mathbf{Y}_p$ as

$$\mathbf{Y}_p = b_1\mathbf{Y}_{r-1}+b_2\mathbf{Y}_{r-2}+\ldots+b_r\mathbf{Y}_0$$

and we can clearly use the same argument as above to show that $\mathbf{Y}_{p+1}$ is linearly dependent on the vectors $\mathbf{Y}_0, \mathbf{Y}_1, \ldots, \mathbf{Y}_{r-1}$. Hence by induction the theorem is proved.

From equation (4.2) we get

$$\mathbf{A}^r\mathbf{Y}_0 = a_1\mathbf{A}^{r-1}\mathbf{Y}_0+a_2\mathbf{A}^{r-2}\mathbf{Y}_0+\ldots+a_{r-1}\mathbf{A}\mathbf{Y}_0+a_r\mathbf{Y}_0$$

or

$$(\mathbf{A}^r-a_1\mathbf{A}^{r-1}-a_2\mathbf{A}^{r-2}-\ldots-a_{r-1}\mathbf{A}-a_r\mathbf{I})\mathbf{Y}_0 = 0$$

which we can write as

$$g(\mathbf{A})\mathbf{Y}_0 = 0$$

where

$$g(\lambda) = \lambda^r-a_1\lambda^{r-1}-a_2\lambda^{r-2}-\ldots-a_{r-1}\lambda-a_r$$

Now $g(\lambda)$ is called the *minimal polynomial* of $\mathbf{Y}_0$ with respect to the matrix $\mathbf{A}$, and r is called the *grade* of $\mathbf{Y}_0$ with respect to $\mathbf{A}$.

We shall now show that $g(\lambda)$ is a factor of the minimal polynomial of $\mathbf{A}$, which in turn is a factor of the characteristic equation of $\mathbf{A}$.

Theorem 4.2

If $h(\lambda)$ is the minimal polynomial of $\mathbf{A}$, that is $h(\lambda)$ is the polynomial of least degree such that $h(\mathbf{A}) = 0$, then there exists a polynomial $q(\lambda)$ such that

$$h(\lambda) = g(\lambda)\,q(\lambda)$$

Proof

We first note that since $h(\mathbf{A}) = 0$ obviously $h(\mathbf{A})\mathbf{Y}_0 = 0$. By the division algorithm for polynomials there exist unique polynomials $q(\lambda)$ and $s(\lambda)$ such that

$$h(\lambda) = g(\lambda)\,q(\lambda)+s(\lambda)$$

where $s(\lambda)$ is of lesser degree than $g(\lambda)$. But $h(\mathbf{A})\mathbf{Y}_0 = 0$ and $g(\mathbf{A})\mathbf{Y}_0 = 0$, hence $s(\mathbf{A})\mathbf{Y}_0 = 0$. This means that $s(\mathbf{A})$ must be null for otherwise $g(\mathbf{A})$

is not the minimal polynomial of $\mathbf{Y}_0$ with respect to $\mathbf{A}$. Hence $s(\lambda)$ is null and

$$h(\lambda) = g(\lambda)\, q(\lambda)$$

as required.

It is of course possible to have $h(\lambda) = g(\lambda)$.

Theorem 4.3

If $h(\lambda)$ is the minimal polynomial of $\mathbf{A}$ and $f(\lambda)$ is the characteristic equation of $\mathbf{A}$, then there exists a polynomial $q(\lambda)$ such that

$$f(\lambda) = h(\lambda)\, q(\lambda)$$

Proof

By the division algorithm for polynomials there exist unique polynomials $q(\lambda)$ and $s(\lambda)$ such that

$$f(\lambda) = h(\lambda)\, q(\lambda) + s(\lambda)$$

where $s(\lambda)$ is of degree less than that of $h(\lambda)$. But $f(\mathbf{A}) = 0$ and $h(\mathbf{A}) = 0$, hence $s(\mathbf{A}) = 0$. This means that $s(\mathbf{A})$ must be null for otherwise $h(\mathbf{A})$ would not be the minimal polynomial of $\mathbf{A}$. Hence $s(\lambda)$ is null and

$$f(\lambda) = h(\lambda)\, q(\lambda)$$

as required.

It is of course possible to have $f(\lambda) = h(\lambda)$.

We can now see that by solving equation (4.2) we find the coefficients of the minimal polynomial of $\mathbf{Y}_0$ with respect to $\mathbf{A}$. If we solve this polynomial we find some, or all, of the latent roots of $\mathbf{A}$. The main difficulty in the method is that we are unlikely to know in advance the grade of the vector $\mathbf{Y}_0$. Note that any latent vector of $\mathbf{A}$ has grade 1. A good computing scheme for Krylov is given by both Berezin and Zhidkov,† and by Gantmacher‡ which deals well with the above mentioned difficulty. Gantmacher gives a full discussion of Krylov's method. The scheme proposed by Fadeeva§ is inefficient by comparison because, even if $r < n$, we have to find n vectors before determining the value of r, whereas in the above-mentioned scheme we only have to determine r vectors. The execution of Krylov's method is not discussed here because, as we shall see, it is effectively the same as Danilevsky's method, but whereas Danilevsky's method always allows us to find the characteristic equation, in the case when $r < n$ Krylov's method does not yield the characteristic equation.

† See reference 8, p. 190.
‡ See reference 10, Vol. 1, pp. 202–214.
§ See reference 6, p. 158.

4.2 Relationship Between the Krylov and Danilevsky Methods

Let $\mathbf{F}$ be the Frobenius matrix given by

$$\mathbf{F} = \begin{pmatrix} 0 & 0 & 0 & \dots & 0 & a_r \\ 1 & 0 & 0 & \dots & 0 & a_{r-1} \\ 0 & 1 & 0 & \dots & 0 & a_{r-2} \\ \vdots & \vdots & \vdots & & \vdots & \vdots \\ 0 & 0 & 0 & \dots & 1 & a_1 \end{pmatrix}$$

Although this is of slightly different form to the Frobenius matrix considered in § 3.1, it still retains the important property that the characteristic equation of $\mathbf{F}$ is given by

$$\lambda^r - a_1\lambda^{r-1} - a_2\lambda^{r-2} - \dots - a_{r-1}\lambda - a_r = 0$$

and for this reason is also called a Frobenius matrix. We also let $\mathbf{Y}$ be the matrix whose columns are the vectors $\mathbf{Y}_0, \mathbf{Y}_1, \dots, \mathbf{Y}_{r-1}$ and y_{ij} be the jth component of $\mathbf{Y}_i$. We shall now show that

$$\mathbf{AY} = \mathbf{YF}$$

Firstly,

$$\mathbf{AY} = \mathbf{A}[\mathbf{Y}_0\mathbf{Y}_1 \dots \mathbf{Y}_{r-1}] = [\mathbf{AY}_0\mathbf{AY}_1 \dots \mathbf{AY}_{r-1}]$$

but from equation (4.1) this gives

$$\mathbf{AY} = [\mathbf{Y}_1\mathbf{Y}_2 \dots \mathbf{Y}_r]$$

Also

$$\mathbf{YF} = \begin{pmatrix} y_{01} & y_{11} & y_{21} & \cdots & y_{r-2,1} & y_{r-1,1} \\ y_{02} & y_{12} & y_{22} & \cdots & y_{r-2,2} & y_{r-1,2} \\ y_{03} & y_{13} & y_{23} & \cdots & y_{r-2,3} & y_{r-1,3} \\ \vdots & \vdots & \vdots & & \vdots & \vdots \\ y_{0n} & y_{1n} & y_{2n} & \cdots & y_{r-2,n} & y_{r-1,n} \end{pmatrix} \begin{pmatrix} 0 & 0 & 0 & \dots & 0 & a_r \\ 1 & 0 & 0 & \dots & 0 & a_{r-1} \\ 0 & 1 & 0 & \dots & 0 & a_{r-2} \\ \vdots & \vdots & \vdots & & \vdots & \vdots \\ 0 & 0 & 0 & \dots & 1 & a_1 \end{pmatrix}$$

$$= \begin{pmatrix} y_{11} & y_{21} & y_{31} & \cdots & y_{r-1,1} & (a_r y_{01} + a_{r-1}y_{11} + a_{r-2}y_{21} + \dots + a_1 y_{r-1,1}) \\ y_{12} & y_{22} & y_{32} & \cdots & y_{r-1,2} & (a_r y_{02} + a_{r-1}y_{12} + a_{r-2}y_{22} + \dots + a_1 y_{r-1,2}) \\ y_{13} & y_{23} & y_{33} & \cdots & y_{r-1,3} & (a_r y_{03} + a_{r-1}y_{13} + a_{r-2}y_{23} + \dots + a_1 y_{r-1,3}) \\ \vdots & \vdots & \vdots & & \vdots & \vdots \\ y_{1n} & y_{2n} & y_{3n} & \cdots & y_{r-1,n} & (a_r y_{0n} + a_{r-1}y_{1n} + a_{r-2}y_{2n} + \dots + a_1 y_{r-1,n}) \end{pmatrix}$$

From equation (4.2) the last column is $\mathbf{Y}_r$ so that

$$\mathbf{YF} = [\mathbf{Y}_1\mathbf{Y}_2 \dots \mathbf{Y}_r] = \mathbf{AY}$$

In the special case $r = n$, $\mathbf{Y}^{-1}$ exists so that

$$\mathbf{F} = \mathbf{Y}^{-1}\mathbf{A}\mathbf{Y}$$

which is of course the similarity transformation produced by Danilevsky's method.

So we can see that Krylov's method is also indirectly attempting to find the companion matrix of **A**. Whereas Danilevsky's method always allows us to find the characteristic equation, when $r < n$ Krylov's method only yields a factor of the characteristic equation.†

Example 4.1

Here we take the matrix **A** of example 3.1, that is

$$\mathbf{A} = \begin{pmatrix} 0 & -2 & 5 \\ -7 & 1 & 9 \\ -1 & -2 & 6 \end{pmatrix} \quad \text{and also we put } \mathbf{Y}_0 = \begin{pmatrix} 1 \\ 0 \\ 0 \end{pmatrix}$$

so that

$$\mathbf{Y}_1 = \mathbf{A}\mathbf{Y}_0 = \begin{pmatrix} 0 \\ -7 \\ -1 \end{pmatrix}, \quad \mathbf{Y}_2 = \mathbf{A}\mathbf{Y}_1 = \begin{pmatrix} 9 \\ -16 \\ 8 \end{pmatrix}, \quad \mathbf{Y}_3 = \mathbf{A}\mathbf{Y}_2 = \begin{pmatrix} 72 \\ -7 \\ 71 \end{pmatrix}$$

Here it is easily seen that $\mathbf{Y}_0$, $\mathbf{Y}_1$ and $\mathbf{Y}_2$ are linearly independent so that the minimal polynomial of $\mathbf{Y}_0$ with respect to **A** is in fact the characteristic equation of **A**. From

$$\mathbf{Y}_3 = a_1\mathbf{Y}_2 + a_2\mathbf{Y}_1 + a_3\mathbf{Y}_0$$

we get the equations

$$\begin{aligned} a_3 + \qquad 9a_1 &= 72 \\ -7a_2 - 16a_1 &= -7 \\ -\ a_2 + \ 8a_1 &= 71 \end{aligned}$$

which have the solution

$$a_1 = 7, \quad a_2 = -15, \quad a_3 = 9$$

so that the characteristic equation of **A** is

$$\lambda^3 - 7\lambda^2 + 15\lambda - 9 = 0$$

Krylov's method, in this example, has given the characteristic equation of **A**. We note that Danilevsky's method gave the complete Frobenius form.

† Theorem 4.1 shows that we can only find r coefficients.

Example 4.2

Here we take the matrix $\mathbf{A}$ of example 3.3, so that

$$\mathbf{A} = \begin{pmatrix} 1 & -1 & 3 & 4 \\ 4 & 1 & 2 & 1 \\ 4 & 2 & 1 & -1 \\ 0 & -1 & 1 & 0 \end{pmatrix}$$

and put

$$\mathbf{Y}_0 = \begin{pmatrix} 1 \\ 0 \\ 0 \\ 0 \end{pmatrix}$$

so that

$$\mathbf{Y}_1 = \mathbf{A}\mathbf{Y}_0 = \begin{pmatrix} 1 \\ 4 \\ 4 \\ 0 \end{pmatrix} \quad \text{and} \quad \mathbf{Y}_2 = \mathbf{A}\mathbf{Y}_1 = \begin{pmatrix} 9 \\ 16 \\ 16 \\ 0 \end{pmatrix}$$

It is easily verified that

$$\mathbf{Y}_2 = 4\mathbf{Y}_1 + 5\mathbf{Y}_0$$

so that the grade of $\mathbf{Y}_0$ with respect to $\mathbf{A}$ is only 2 and Krylov's method only yields the equation

$$\lambda^2 - 4\lambda - 5 = 0$$

We note that this is the characteristic equation of the matrix $\mathbf{D}_1$ of Danilevsky's method.

This example is of some interest, for being non-derogatory, $\mathbf{A}$ is similar to its companion matrix† and yet neither Danilevsky's method nor Krylov's method with this starting vector yields this.

4.3 Further Comments on Krylov's Method

Example 4.2 illustrates well the shortcomings of Krylov's method, for not only have we not found the full characteristic equation, but we have not even found all the distinct latent roots of $\mathbf{A}$. In certain cases a different starting vector may yield the characteristic equation,‡ but the uncertainty makes the method of little practical value. Furthermore, if $r < n$ it is not easy to find the latent vector by Krylov's method.

† See reference 7, p. 13.
‡ See exercise 4.2.

4.4 Exercises

4.1. Apply Krylov's method to the matrices of exercises 3.1 (i) and (ii).

4.2. Apply Krylov's method to the matrix **A** of example 4.2 taking

$$\mathbf{Y}_0 = \begin{pmatrix} 0 \\ 1 \\ 0 \\ 0 \end{pmatrix}$$

4.3. Show how the latent vectors may be found in the case where Krylov's method has given the characteristic equation. Why cannot this method be used when $r < n$?

4.4. If **A** is tridiagonal (see § 5.3) and $\mathbf{Y}_0^T = (1 \quad 0 \quad 0 \quad \dots \quad 0)$, show that the matrix **Y** of § 4.2 is upper triangular. Hence show that Krylov's method gives the characteristic equation if, and only if, the elements on the diagonal below the leading diagonal are non-zero.

Chapter 5

FINDING THE LATENT ROOTS OF A TRIDIAGONAL MATRIX

One important group of methods of finding the latent roots of a matrix, **A**, involves obtaining a tridiagonal matrix which is similar to **A**. The latent roots of many tridiagonal matrices, in particular symmetric ones, can be located using the properties of a Sturm series. For this reason we shall first outline the theory for Sturm sequences.

5.1 Sturm Series and Sturm's Theorem

If we have a sequence of polynomials,

$$f_n(x), f_{n-1}(x), \ldots, f_1(x), f_0(x)$$

which satisfy the following three (sufficient) conditions,† then the sequence is called a Sturm series for $f_n(x)$.

1. When x increases through a real root of $f_n(x)$, the product $f_n(x) . f_{n-1}(x)$ changes sign, either always from + to −, or always from − to +.

2. If when $x = a$, $f_r(a) = 0$, then $f_{r+1}(a)$ and $f_{r-1}(a)$ have opposite signs. (Hence neither is zero.)

3. $f_0(x)$ does not have real roots.

The sequence is called a Sturm series for the polynomial $f_n(x)$ in (a, b) if the above three properties hold in this interval.

Theorem 5.1 (*Sturm's theorem*)

If we have a Sturm series for $f_n(x)$ in (a, b), a and b not being roots of $f_n(x)$, then the number of distinct real roots of the polynomial $f_n(x)$ in (a, b) is given by $|S(a) - S(b)|$, where $S(\alpha)$ is the number of changes of sign in the sequence $f_n(\alpha), f_{n-1}(\alpha), \ldots, f_1(\alpha), f_0(\alpha)$.

Proof

As x increases through a root of $f_r(x)$, where $1 \leqslant r \leqslant n-1$, then by condition (2) the value of $S(x)$ does not alter. By (3) this is also true when $r = 0$. On the other hand, as x increases through a root of $f_n(x)$, then by (1) $S(x)$ either always increases by one or always decreases by one. The theorem now follows.

† Many variations are to be found. This seems to have arisen because Sturm's theorem was originally proved with only a particular sequence in mind. As far as I know necessary conditions have not been proposed.

5.2 Construction of a Sturm Series

Suppose we have a polynomial $f(x) = 0$ having distinct roots. We form the following equations

$$f_n(x) = f(x), \quad f_{n-1}(x) = f'(x)$$

and

$$\begin{aligned} f_n(x) &= f_{n-1}(x)\, q_1(x) - f_{n-2}(x) \\ f_{n-1}(x) &= f_{n-2}(x)\, q_2(x) - f_{n-3}(x) \\ \vdots \quad & \quad \vdots \quad \quad \vdots \\ f_2(x) &= f_1(x)\, q_{n-1}(x) - f_0(x) \end{aligned}$$

where $f_r(x)$ is of lower degree than $f_{r+1}(x)$. The sequence of polynomials $f_n(x), f_{n-1}(x), \ldots, f_0(x)$ can easily be shown to satisfy the conditions of a Sturm series.† The above is the classical construction of the Sturm series, and it is perhaps a misnomer to call other sequences Sturm series.

Example 5.1

To locate the root in $(-1, 1)$ of the polynomial,

$$f(x) = x^4 - 3x^3 - x^2 + 8x - 4$$

Using the above construction we get

$$\begin{aligned} f_4(x) &= f(x) = x^4 - 3x^3 - x^2 + 8x - 4 \\ f_3(x) &= f'(x) = 4x^3 - 9x^2 - 2x + 8 \\ f_2(x) &= \tfrac{1}{16}(35x^2 - 90x + 40) = \tfrac{5}{16}(7x^2 - 18x + 8) \\ f_1(x) &= \tfrac{1}{49}(160x - 320) = \tfrac{160}{49}(x - 2) \\ f_0(x) &= 0 \end{aligned}$$

x	f_4	f_3	f_2	f_1	f_0	$S(x)$	Comment
-1	$-$	$-$	$+$	$-$		2	
1	$+$	$+$	$-$	$-$		1	One root in $(-1, 1)$
0	$-$	$+$	$+$	$-$		2	One root in $(0, 1)$
0·5	$-$	$+$	$+$	$-$		2	One root in $(0{\cdot}5, 1)$
0·75	$+$	$+$	$-$	$-$		1	One root in $(0{\cdot}5, 0{\cdot}75)$

Clearly we may continue this process to achieve any desired accuracy.

† See reference 11, p. 199.

5.3 Sturm's Theorem and the Latent Roots of a Tridiagonal Matrix

We wish to find the latent roots of the matrix

$$\mathbf{A} = \begin{pmatrix} a_1 & c_2 & 0 & 0 & \dots & 0 & 0 & 0 \\ b_2 & a_2 & c_3 & 0 & \dots & 0 & 0 & 0 \\ 0 & b_3 & a_3 & c_4 & \dots & 0 & 0 & 0 \\ \vdots & \vdots & \vdots & \vdots & & \vdots & \vdots & \vdots \\ 0 & 0 & 0 & 0 & \dots & 0 & b_n & a_n \end{pmatrix}$$

We have to solve the determinantal equation

$$f_n(\lambda) = |\mathbf{A} - \lambda\mathbf{I}| = 0$$

Let $f_r(\lambda)$ be the determinant formed by the first r rows and columns of $f_n(\lambda)$ so that

$$f_r(\lambda) = \begin{vmatrix} a_1-\lambda & c_2 & 0 & \dots & 0 & 0 & 0 \\ b_2 & a_2-\lambda & c_3 & \dots & 0 & 0 & 0 \\ \vdots & \vdots & \vdots & & \vdots & \vdots & \vdots \\ 0 & 0 & 0 & \dots & b_{r-1} & a_{r-1}-\lambda & c_r \\ 0 & 0 & 0 & \dots & 0 & b_r & a_r-\lambda \end{vmatrix}$$

Expanding $f_r(\lambda)$ by the last row we get

$$f_r(\lambda) = (a_r - \lambda)f_{r-1}(\lambda) - b_r c_r f_{r-2}(\lambda) \tag{5.1}$$

where $f_0(\lambda) = 1$ and $f_1(\lambda) = a_1 - \lambda$.

We shall show that the sequence $f_n(\lambda), f_{n-1}(\lambda), \dots, f_0(\lambda)$ is a Sturm series in $f_n(\lambda)$, the characteristic equation of $\mathbf{A}$, providing that for all r we have $b_r c_r > 0$. In particular this is true when $\mathbf{A}$ is symmetric with $b_r \neq 0$ for all r.

In order to obtain this important result we require some preliminary theorems.

Theorem 5.2

The tridiagonal matrix $\mathbf{A}$ with $b_r c_r > 0$ for all r, is similar to a symmetric tridiagonal matrix having non-zero superdiagonal elements.

Proof

We prove the theorem by showing that there exists a diagonal matrix $\mathbf{D}$ such that $\mathbf{B} = \mathbf{D}\mathbf{A}\mathbf{D}^{-1}$, where $\mathbf{B}$ is the required symmetric matrix. Let

$$\mathbf{DAD}^{-1} = \begin{pmatrix} d_1 & 0 & 0 & \dots & 0 \\ 0 & d_2 & 0 & \dots & 0 \\ 0 & 0 & d_3 & \dots & 0 \\ \vdots & \vdots & \vdots & & \vdots \\ 0 & 0 & 0 & \dots & d_n \end{pmatrix} \begin{pmatrix} a_1 & c_2 & 0 & \dots & 0 \\ b_2 & a_2 & c_3 & \dots & 0 \\ 0 & b_3 & a_3 & \dots & 0 \\ \vdots & \vdots & \vdots & & \vdots \\ 0 & 0 & 0 & \dots & a_n \end{pmatrix} \begin{pmatrix} \frac{1}{d_1} & 0 & 0 & \dots & 0 \\ 0 & \frac{1}{d_2} & 0 & \dots & 0 \\ 0 & 0 & \frac{1}{d_3} & \dots & 0 \\ \vdots & \vdots & \vdots & & \vdots \\ 0 & 0 & 0 & \dots & \frac{1}{d_n} \end{pmatrix}$$

$$= \begin{pmatrix} a_1 d_1 & c_2 d_1 & 0 & \dots & 0 \\ b_2 d_2 & a_2 d_2 & c_3 d_2 & \dots & 0 \\ 0 & b_3 d_3 & a_3 d_3 & \dots & 0 \\ \vdots & \vdots & \vdots & & \vdots \\ 0 & 0 & 0 & \dots & a_n d_n \end{pmatrix} \begin{pmatrix} \frac{1}{d_1} & 0 & 0 & \dots & 0 \\ 0 & \frac{1}{d_2} & 0 & \dots & 0 \\ 0 & 0 & \frac{1}{d_3} & \dots & 0 \\ \vdots & \vdots & \vdots & & \vdots \\ 0 & 0 & 0 & \dots & \frac{1}{d_n} \end{pmatrix}$$

$$= \begin{pmatrix} a_1 & c_2 \cdot \frac{d_1}{d_2} & 0 & \dots & 0 \\ b_2 \cdot \frac{d_2}{d_1} & a_2 & c_3 \cdot \frac{d_2}{d_3} & \dots & 0 \\ 0 & b_3 \cdot \frac{d_3}{d_2} & a_3 & \dots & 0 \\ \vdots & \vdots & \vdots & & \vdots \\ 0 & 0 & 0 & \dots & a_n \end{pmatrix} = \mathbf{B}$$

If **B** is to be symmetric it is clear that we require

$$\frac{d_1^2}{d_2^2} = \frac{b_2}{c_2}, \quad \frac{d_2^2}{d_3^2} = \frac{b_3}{c_3}, \quad \frac{d_3^2}{d_4^2} = \frac{b_4}{c_4}, \quad \dots, \quad \frac{d_{n-1}^2}{d_n^2} = \frac{b_n}{c_n}$$

Since $b_r c_r > 0$, we have that $b_r/c_r > 0$ and hence we are able to select the d_r's so that **B** is symmetric with non-zero superdiagonal elements, as required. This proves the theorem.

Theorem 5.3

No two neighbouring polynomials in the sequence $f_n(\lambda), f_{n-1}(\lambda), \ldots, f_0(\lambda)$ can have a root in common if $b_r c_r > 0$ for all r.

Proof

First we note that $f_0(\lambda)$ and $f_1(\lambda)$ do not have a common root.

Assume that $f_{r-2}(\lambda)$ and $f_{r-1}(\lambda)$ do not have a common root. Now equation (5.1) is

$$f_r(\lambda) = (a_r - \lambda) f_{r-1}(\lambda) - b_r c_r f_{r-2}(\lambda)$$

from which we see that, if our inductive hypothesis holds, $f_r(\lambda) \neq 0$ when $f_{r-1}(\lambda) = 0$, for this would mean otherwise that $f_{r-2}(\lambda) = 0$ also because $b_r c_r \neq 0$. Hence by induction the theorem is proved.

Theorem 5.4

If $\mathbf{B}$ is a symmetric tridiagonal matrix such that none of its superdiagonal elements is zero, then between any two roots of $f_n(\lambda)$ there is a root of $f_{n-1}(\lambda)$.

Proof

$$f_1(\lambda) = a_1 - \lambda$$

which has the root $\lambda = a_1$

$$f_2(\lambda) = (a_2 - \lambda)(a_1 - \lambda) - b_2^2$$

when $\lambda = a_1$,

$$f_2(\lambda) = -b_2^2 < 0$$

when $\lambda = -\infty$,

$$f_2(\lambda) > 0$$

and when $\lambda = \infty$,

$$f_2(\lambda) > 0$$

Hence $f_2(\lambda)$ has one root less than a_1 and one root greater than a_1.

Now assume that $f_{n-1}(\lambda)$ has $n-1$ distinct roots and that between each of these there is a root of $f_{n-2}(\lambda)$.

Suppose that two neighbouring roots of $f_{n-1}(\lambda)$ are λ_1 and λ_2, with $\lambda_1 < \lambda_2$. Now,

$$f_n(\lambda) = (a_n - \lambda) f_{n-1}(\lambda) - b_n^2 f_{n-2}(\lambda)$$

Hence

$$f_n(\lambda_1) = -b_n^2 f_{n-2}(\lambda_1)$$

and

$$f_n(\lambda_2) = -b_n^2 f_{n-2}(\lambda_2)$$

But by the above assumption $f_{n-2}(\lambda)$ changes sign in (λ_1, λ_2) and hence $f_n(\lambda)$ must also change sign. It follows that between any two roots of $f_{n-1}(\lambda)$ there is a root of $f_n(\lambda)$.

By considering the limit as $\lambda \to -\infty$, it is easy to show that $f_r(\lambda) > 0$ for all r and hence $f_n(\lambda)$ has a root less than the smallest root of $f_{n-1}(\lambda)$. Its remaining root can only be larger than the largest root of $f_{n-1}(\lambda)$. By induction the theorem is now proved.

Notice that we have also established that the roots of $f_n(\lambda)$ are all distinct.

Theorem 5.5

If **A** is a tridiagonal matrix with $b_r c_r > 0$ for all r, then between any two roots of $f_n(\lambda)$ there is a root of $f_{n-1}(\lambda)$.

Proof

In theorem 5.2 we can clearly choose **D** so that the principal submatrix of **A** is similar to the principal submatrix of **B**. The result now follows from theorem 5.3.

We are now in a position to prove our main results.

Theorem 5.6

The sequence $f_n(\lambda), f_{n-1}(\lambda), \ldots, f_0(\lambda)$ is a Sturm series in $f_n(\lambda)$ providing that $b_r c_r > 0$ for all r.

Proof

We now refer to the conditions of § 5.1. Condition (1) follows immediately from theorem 5.4, condition (2) from theorem 5.2 and equation (5.1) since $b_r c_r > 0$, and condition (3) is satisfied since $f_0(\lambda) = 1$. Hence the sequence is a Sturm series and the theorem is proved.

This, of course, gives us a powerful method for locating the latent roots of this type of tridiagonal matrix. We shall now show that this particular sequence is even more convenient than the general Sturm series.

Theorem 5.7

For the above Sturm series, if α is not a root of $f_n(\lambda)$, then $\mathbf{S}(\alpha)$ is the number of roots of $f_n(\lambda)$ less than α.

Proof

When $\lambda = -\infty$, $f_r(\lambda)$ is positive for all r, hence $\mathbf{S}(-\infty) = 0$. So by Sturm's theorem the number of roots in the interval $(-\infty, \alpha)$ is given by

$$|\mathbf{S}(\alpha) - \mathbf{S}(-\infty)| = \mathbf{S}(\alpha)$$

and the theorem is proved.

We note that having located a root using Sturm's theorem we are likely to obtain much better convergence using a method such as Newton's approximation. Equation (5.1) gives us a useful relationship for applying Newton's approximation since from

$$f_r(\lambda) = (a_r - \lambda) f_{r-1}(\lambda) - b_r c_r f_{r-2}(\lambda)$$

we get

$$f'_r(\lambda) = (a_r - \lambda) f'_{r-1}(\lambda) - f_{r-1}(\lambda) - b_r c_r f'_{r-2}(\lambda) \tag{5.2}$$

where $f'_0(\lambda) = 0$ and $f'_1(\lambda) = -1$.

Example 5.2

To find the middle latent root of

$$\mathbf{A} = \begin{pmatrix} 1 & -1 & 0 \\ -1 & 2 & 1 \\ 0 & 2 & 3 \end{pmatrix}$$

By Gerschgorin's theorem all the latent roots of **A** are in the interval $(-5, 5)$. Now,

$$f_0(\lambda) = 1$$

$$f_1(\lambda) = 1 - \lambda$$

$$f_2(\lambda) = (2 - \lambda) f_1(\lambda) - 1$$

$$f_3(\lambda) = (3 - \lambda) f_2(\lambda) - 2 f_1(\lambda)$$

x	f	f	f	f	$\mathrm{S}(x)$	Comment
-5	$+1$	$+6$	$+41$	$+$	0	No roots < -5 (as expected)
5	$+1$	-4	$+11$	$-$	3	3 roots in $(-5, 5)$ (also as expected)
0	$+1$	$+1$	$+1$	$+$	0	3 roots in $(0, 5)$
2·5	$+1$	$-1{\cdot}5$	$-0{\cdot}25$	$+$	2	2 roots in $(0, 2{\cdot}5)$, 1 in $(2{\cdot}5, 5)$
1·25	$+1$	$-0{\cdot}25$	$-1{\cdot}1875$	$-$	1	Middle root in $(1{\cdot}25, 2{\cdot}5)$

We have now determined that there is one root in each of the intervals $(0, 1{\cdot}25)$, $(1{\cdot}25, 2{\cdot}5)$, $(2{\cdot}5, 5)$. We may continue the process of bisection to find a particular root to any required degree of accuracy. Clearly we improve the root by one binary position at each bisection.

We shall use Newton's approximation to improve the middle root.

Now

$$f_0'(\lambda) = 0$$
$$f_1'(\lambda) = -1$$
$$f_2'(\lambda) = (\lambda-2)-f_1(\lambda)$$
$$f_3'(\lambda) = (3-\lambda)f_2'(\lambda)-f_2(\lambda)+2$$

Taking $x = 1{\cdot}6$ we get

$$f_1 = -0{\cdot}6, \quad f_2 = -1{\cdot}24, \quad f_3 = -0{\cdot}536$$

and

$$f_2' = 0{\cdot}2, \quad f_3' = 3{\cdot}52$$

so that

$$\lambda \simeq 1{\cdot}6+\frac{0{\cdot}536}{3{\cdot}52} = 1{\cdot}6+0{\cdot}15 = 1{\cdot}75$$

Taking $x = 1{\cdot}75$ we get

$$f_1 = -0{\cdot}75, \quad f_2 = -1{\cdot}1875, \quad f_3 = 0{\cdot}0156$$

and

$$f_2' = 0{\cdot}5, \quad f_3' = 3{\cdot}8125$$

so that

$$\lambda \simeq 1{\cdot}75-\frac{0{\cdot}0156}{3{\cdot}8125} = 1{\cdot}75-0{\cdot}0041 = 1{\cdot}7459$$

Taking $x = 1{\cdot}7459$ we get

$$f_1 = -0{\cdot}7459, \quad f_2 = -1{\cdot}18953319, \quad f_3 = 0{\cdot}00000643$$

and

$$f_2' = 0{\cdot}4918, \quad f_3' = 3{\cdot}8063$$

so that

$$\lambda \simeq 1{\cdot}7459-\frac{0{\cdot}00000643}{3{\cdot}8063} = 1{\cdot}7459-0{\cdot}0000017$$
$$\simeq 1{\cdot}745898$$

Correct to six decimal places we now have $\lambda = 1{\cdot}745898$.

We can see that at the expense of some extra calculation Newton's approximation has given much better convergence than continued bisection, which would have required about twenty iterations to achieve the degree of accuracy obtained here.

If we have $b_r c_r \geqslant 0$ for all r, although the sequence $f_n(\lambda), f_{n-1}(\lambda), \ldots, f_0(\lambda)$ is not a Sturm series we may partition the matrix into two or more tridiagonal submatrices and apply the Sturm theory to these submatrices.

Example 5.3

$$\mathbf{A} = \begin{pmatrix} 1 & 2 & 0 & 0 & 0 & 0 & 0 \\ 3 & -1 & -4 & 0 & 0 & 0 & 0 \\ 0 & -2 & 4 & 0 & 0 & 0 & 0 \\ 0 & 0 & 1 & -3 & 3 & 0 & 0 \\ 0 & 0 & 0 & 5 & 8 & -1 & 0 \\ 0 & 0 & 0 & 0 & -6 & -4 & 1 \\ 0 & 0 & 0 & 0 & 0 & 5 & 3 \end{pmatrix}$$

Here we have $b_4 = 0$ so that $b_4 c_4 = 0$, but for all $r \neq 4$, $b_r c_r > 0$. Because $b_4 = 0$ we have from theorem 1.14 that

$$|\mathbf{A} - \lambda\mathbf{I}| = |\mathbf{D}_1 - \lambda\mathbf{I}|\,|\mathbf{D}_2 - \lambda\mathbf{I}|$$

where

$$\mathbf{D}_1 = \begin{pmatrix} 1 & 2 & 0 \\ 3 & -1 & -4 \\ 0 & -2 & 4 \end{pmatrix} \quad \text{and} \quad \mathbf{D}_2 = \begin{pmatrix} -3 & 3 & 0 & 0 \\ 5 & 8 & -1 & 0 \\ 0 & -6 & -4 & 1 \\ 0 & 0 & 5 & 3 \end{pmatrix}$$

So we may find the latent roots of **B** by applying the Sturm theory to $\mathbf{D}_1$ and $\mathbf{D}_2$ individually. This case clearly simplifies the work involved.

5.4 The Method of Muller

If $b_r c_r < 0$ for any r then we cannot apply the Sturm theory at all. Clearly this case is not possible for a symmetric matrix. Here of course it is harder to locate particular roots unless we have prior knowledge as to their distribution. Having located a root we can of course use Newton's method as discussed in the previous section. For complex roots we can use a method such as Bairstow's method.

A convenient method for finding latent roots in this case, however, is the method of Muller which enables us to use the recurrence relation of equation (5.1) rather than to find the characteristic equation explicitly.

The method of Muller fits a quadratic equation as an approximation to the characteristic equation in the neighbourhood of a root, and takes one root of the quadratic as the approximation to this root. With this approximation to the root we proceed to fit a new quadratic, and so on.

If $f(\lambda)$ is the characteristic equation, we take three points x_1, x_2 and x_3 in the neighbourhood of a root and fit a quadratic through the points $[x_1,f(x_1)]$, $[x_2,f(x_2)]$ and $[x_3,f(x_3)]$ and then replace one of the x's by one of the roots of the quadratic and repeat this process.

Example 5.4

To find the latent roots of the matrix

$$\mathbf{A} = \begin{pmatrix} 0 & 9 & 0 \\ 1 & 8 & -32 \\ 0 & 1 & -1 \end{pmatrix}$$

Since $b_3 c_3 = -32$ we cannot use Sturm's theory.

Now

$$f_0(\lambda) = 1$$

$$f_1(\lambda) = -\lambda$$

$$f_2(\lambda) = (8-\lambda)f_1(\lambda) - 9$$

$$f_3(\lambda) = -(1+\lambda)f_2(\lambda) + 32f_1(\lambda)$$

Let us take as our initial points $x_1 = -2$, $x_2 = 0$, $x_3 = 2$. At x_1,

$$f_1 = 2, \quad f_2 = 11, \quad f_3 = 75$$

at x_2,

$$f_1 = 0, \quad f_2 = -9, \quad f_3 = 9$$

at x_3,

$$f_1 = -2, \quad f_2 = -21, \quad f_3 = -1$$

(Since f_3 has changed sign there is a root in $(0, 2)$.) We wish to fit a quadratic $y = ax^2+bx+c$ through the points $(-2, 75)$, $(0, 9)$ and $(2, -1)$. This gives

$$y = 7x^2 - 19x + 9$$

and putting $y = 0$ we have

$$x \simeq 2{\cdot}1 \quad \text{or} \quad 0{\cdot}61$$

Since we have established that there is a root between 0 and 2 we replace -2 by $0{\cdot}61$, and take as our new points $x_1 = 0$, $x_2 = 0{\cdot}61$, $x_3 = 2$.

At x_2,

$$f_1 = -0{\cdot}61, \quad f_2 = -13{\cdot}5079, \quad f_3 \simeq 2{\cdot}228$$

(The root is in $(0{\cdot}61, 2)$.)

Fitting a quadratic through $(0, 9)$, $(0{\cdot}61, 2{\cdot}23)$, $(2, -1)$ we get

$$y = 4{\cdot}387x^2 - 13{\cdot}775x + 9$$

and putting $y = 0$ we have

$$x \simeq 0{\cdot}93 \quad \text{or} \quad 2{\cdot}2$$

Taking $x_1 = 0$, $x_2 = 0{\cdot}61$, $x_3 = 0{\cdot}93$, at x_3,

$$f_1 = -0{\cdot}93, \quad f_2 = -15{\cdot}5751, \quad f_3 \simeq 0{\cdot}300$$

Fitting a quadratic through $(0, 9)$, $(0{\cdot}61, 2{\cdot}228)$, $(0{\cdot}93, 0{\cdot}300)$ we get

$$y = 5{\cdot}3642x^2 - 14{\cdot}1815x + 9$$

and putting $y = 0$ we have

$$x \simeq 1{\cdot}06 \quad \text{or} \quad 2$$

At $x = 1{\cdot}06$,

$$f_1 = -1{\cdot}06, \quad f_2 = -16{\cdot}3564, \quad f_3 = -0{\cdot}225816$$

Clearly we are now close to the root. It lies in the interval $(0{\cdot}93, 1{\cdot}06)$. We now use Newton's method commencing with $x = 0{\cdot}995$ (the mean of the two values). Now

$$f_1'(\lambda) = -1$$
$$f_2'(\lambda) = (\lambda - 8) - f_1(\lambda)$$
$$f_3'(\lambda) = -(1+\lambda)f_2'(\lambda) - f_2(\lambda) - 32$$

When $x = 0{\cdot}995$,

$$f_1 = -0{\cdot}995, \quad f_2 = -15{\cdot}969955, \quad f_3 = 0{\cdot}020060$$

and

$$f_2' = -6{\cdot}010, \quad f_3' = -5{\cdot}040095$$

so that

$$\lambda \simeq 0{\cdot}995 + \frac{0{\cdot}020060}{5{\cdot}040095} = 0{\cdot}995 + 0{\cdot}00398 = 0{\cdot}99898$$

When $x = 0{\cdot}99898$,

$$f_1 = -0{\cdot}99898, \quad f_2 = -15{\cdot}99387896, \quad f_3 = 0{\cdot}00408416$$

and

$$f_2' = -6{\cdot}00204, \quad f_3' = -4{\cdot}00816312$$

so that

$$\lambda \simeq 0{\cdot}99898 + \frac{0{\cdot}00408416}{4{\cdot}00816312} = 0{\cdot}99898 + 0{\cdot}0010190$$

$$\simeq 0{\cdot}9999990$$

Correct to five decimal places the latent root is

$$\lambda_1 = 1{\cdot}00000$$

which corresponds exactly to the correct latent root.

To find the remaining roots we again use the method of Muller, but we divide $f_3(\lambda)$ by the root we have just found so that we do not converge to this root once more. For this reason it is important to find the root with which we are dividing accurately. Since, in this example, there are only two roots remaining these will of course be the roots of the fitted quadratic. Put

$$g_3(\lambda) = \frac{f_3(\lambda)}{\lambda-\lambda_1} = \frac{f_3(\lambda)}{\lambda-1{\cdot}00000}$$

Again taking $x_1 = -2$, $x_2 = 0$, $x_3 = 2$, we get

$$g_3(-2) = -25, \quad g_3(0) = -9, \quad g_3(2) = -1$$

and the fitted quadratic is

$$y = x^2 - 6x + 9 = (x-3)^2$$

So the latent roots of $\mathbf{A}$ are

$$\lambda_1 = 1{\cdot}00000, \quad \lambda_2 = \lambda_3 = 3{\cdot}00000$$

We can see from this example that it is of some importance to look at the behaviour of the characteristic equation to make sure that we are converging to a root.

In the case of complex roots, Muller's method, of course, involves complex arithmetic. The local convergence, which is proved by Muller,† is generally about 1·8 for single roots and somewhat slower for multiple roots. It is thought to converge globally, which means that arbitrary starting values can be used. It should be borne in mind that roots of polynomials are best found in ascending order of magnitude if the polynomial is to be deflated by that root as in example 5.4.‡ If arbitrary starting values are used in Muller's method, they should be centred around zero in the hope that the smallest root is found first.

5.5 Exercises

5.1. (i) Use Sturm's theorem to locate to an accuracy of $\pm 0{\cdot}5$ the three real roots of the equation

$$9x^3 - 40x - 21 = 0$$

(ii) Use Sturm's theorem to locate to an accuracy of $\pm 0{\cdot}5$ the real roots of the equation,

$$x^4 + 4x^3 + 6x^2 + 4x - 24 = 0$$

5.2. If

$$f_r(\lambda) = (a_r - \lambda) f_{r-1}(\lambda) - b_r^2 f_{r-2}(\lambda)$$

† See reference 9.

‡ See reference 17, Chapter 2, pp. 55–65.

where $f_0(\lambda) = 1$ and $f_1(\lambda) = a_1 - \lambda$ prove that

$$f_r(-\infty) > 0 \quad \text{for all } r$$

5.3. (i) If n is a positive integer greater than one and $1 \leqslant i \leqslant n-1$ show that the angles given by

$$\alpha = \frac{(2i-1)\pi}{2n}, \quad \beta = \frac{(2i-1)\pi}{2(n-1)}, \quad \gamma = \frac{(2i+1)\pi}{2n}$$

satisfy

$$0 < \alpha < \beta < \gamma < \pi$$

(ii) Hence, or otherwise, prove that between any two roots of the Chebyshev polynomial $\mathbf{T}_n(x)$ there is a root of $\mathbf{T}_{n-1}(x)$, where

$$\mathbf{T}_n(x) = \cos n\theta \quad \text{and} \quad \cos\theta = x$$

(iii) Prove that the sequence $\mathbf{T}_n(x), \mathbf{T}_{n-1}(x), \ldots, \mathbf{T}_0(x)$ forms a Sturm series for $\mathbf{T}_n(x)$ in the interval $(-1, 1)$.

5.4. Let $\mathbf{P}_n(x), \mathbf{P}_{n-1}(x), \ldots, \mathbf{P}_0(x)$ be a sequence of orthogonal polynomials, that is,

$$\int_a^b w(x)\,\mathbf{P}_r(x)\,\mathbf{P}_q(x)\,dx = 0 \quad \text{for } r \neq q$$

where $w(x)$ is a weighting function. Prove that the sequence $\mathbf{P}_n(x), \mathbf{P}_{n-1}(x), \ldots, \mathbf{P}_0(x)$ forms a Sturm series for $\mathbf{P}_n(x)$ in the interval (a, b).

5.5. Locate the latent roots of the following tridiagonal matrices

$$\text{(i)}\quad \mathbf{A} = \begin{pmatrix} -1 & 1 & 0 \\ 2 & -3 & -2 \\ 0 & -1 & -5 \end{pmatrix} \qquad \text{(ii)}\quad \mathbf{A} = \begin{pmatrix} 7 & -4 & 0 \\ 11 & -7 & 1 \\ 0 & -1 & 1 \end{pmatrix}$$

5.6. (i) Show that the roots of the quadratic

$$ax^2 + bx + c = 0$$

can be found from the formula

$$x = \frac{-2c}{b \pm \sqrt{(b^2 - 4ac)}}$$

(If $4ac$ is small then the smallest root of the quadratic can be found more accurately from the above formula than the standard formula.)

(ii) Let x_i be the ith approximation to the root of $f(x) = 0$. Also, let

$$\begin{aligned} h_i &= x_i - x_{i-1} \\ \lambda_i &= h_i / h_{i-1} \\ \delta_i &= 1 + \lambda_i \\ g_i &= f(x_{i-2})\,\lambda_i^2 - f(x_{i-1})\,\delta_i^2 + f(x_i)\,(\lambda_i + \delta_i) \end{aligned}$$

Then, the next approximation x_{i+1} is given by

$$x_{i+1} = x_i + h_{i+1}$$

where

$$h_{i+1} = \lambda_{i+1} h_i$$

and

$$\lambda_{i+1} = \frac{-2f(x_i)\,\delta_i}{g_i \pm \sqrt{\{g_i^2 - 4f(x_i)\,\delta_i\,\lambda_i[f(x_{i-2})\,\lambda_i - f(x_{i-1})\,\delta_i + f(x_i)]\}}}$$

the sign being chosen so that the denominator has the greater magnitude.

Show that the above is equivalent to Muller's method. This is the computing scheme proposed by Muller.

5.7. (i) Suppose that $x_i = u + vi$ is an approximation to a root of the polynomial

$$f(x) = a_0 x^n + a_1 x^{n-1} + \ldots + a_n = 0$$

and that $f(u+vi) = p+qi$.

If a quadratic $d_0 x^2 + d_1 x + d_2$ is fitted to the three points,

$$[u-vi, f(u-vi)], \quad [0, f(0)], \quad [u+vi, f(u+vi)]$$

show that we can compute d_0, d_1 and d_2 from the equations

$$d_0 = v(p-a_n) - uq$$

$$d_1 = q(u^2-v^2) - 2uv(p-a_n)$$

$$d_2 = -a_n v(u^2+v^2)$$

If x_{i+1} is a root of the above quadratic, under what conditions is it a closer root of $f(x) = 0$?

(ii) Use the method of (i) to locate the latent roots of the tridiagonal matrix

$$\mathbf{A} = \begin{pmatrix} 0 & -1 & 0 & 0 \\ 1 & 0 & -1 & 0 \\ 0 & 1 & 0 & -1 \\ 0 & 0 & 1 & 0 \end{pmatrix}$$

Chapter 6

THE METHOD OF GIVENS

The method of Givens reduces a symmetric matrix to tridiagonal form by means of a series of orthogonal similarity transformations.

6.1 Orthogonal Matrices

As was discussed in § 1.2, the orthogonal matrix given by

$$\mathbf{Y} = \begin{pmatrix} \cos\theta & -\sin\theta \\ \sin\theta & \cos\theta \end{pmatrix}$$

has the effect of rotating the x, y-axes through an angle $-\theta$. This idea is naturally extended to higher dimensions. For example the matrix given by

$$\mathbf{Y} = \begin{pmatrix} 1 & 0 & 0 & 0 & 0 & 0 \\ 0 & 1 & 0 & 0 & 0 & 0 \\ 0 & 0 & \cos\theta & 0 & -\sin\theta & 0 \\ 0 & 0 & 0 & 1 & 0 & 0 \\ 0 & 0 & \sin\theta & 0 & \cos\theta & 0 \\ 0 & 0 & 0 & 0 & 0 & 1 \end{pmatrix} \begin{matrix} \\ \\ \text{row 3} \\ \\ \text{row 5} \\ \\ \end{matrix}$$

$$\text{column 3} \quad \text{column 5}$$

is a six-dimensional orthogonal matrix that has the effect of rotating the x_3, x_5-axes through an angle $-\theta$. For convenience we shall write the x_3, x_5-plane as the $(3, 5)$-plane and similarly for other planes.

6.2 The Method of Givens

We consider as representative a four by four matrix given by

$$\mathbf{A} = \begin{pmatrix} a_{11} & a_{12} & a_{13} & a_{14} \\ a_{12} & a_{22} & a_{23} & a_{24} \\ a_{13} & a_{23} & a_{33} & a_{34} \\ a_{14} & a_{24} & a_{34} & a_{44} \end{pmatrix}$$

First we wish to make a_{13} zero. In order to achieve this we take the orthogonal matrix that rotates in the $(2, 3)$-plane. Putting $c = \cos\theta$ and

$s = \sin\theta$ we have

$$\mathbf{A}_1 = \mathbf{Y}_1^{-1}\mathbf{A}\mathbf{Y}_1 = \mathbf{Y}_1^T\mathbf{A}\mathbf{Y}_1$$

$$= \begin{pmatrix} 1 & 0 & 0 & 0 \\ 0 & c & s & 0 \\ 0 & -s & c & 0 \\ 0 & 0 & 0 & 1 \end{pmatrix} \begin{pmatrix} a_{11} & a_{12} & a_{13} & a_{14} \\ a_{12} & a_{22} & a_{23} & a_{24} \\ a_{13} & a_{23} & a_{33} & a_{34} \\ a_{14} & a_{24} & a_{34} & a_{44} \end{pmatrix} \begin{pmatrix} 1 & 0 & 0 & 0 \\ 0 & c & -s & 0 \\ 0 & s & c & 0 \\ 0 & 0 & 0 & 1 \end{pmatrix}$$

$$= \begin{pmatrix} 1 & 0 & 0 & 0 \\ 0 & c & s & 0 \\ 0 & -s & c & 0 \\ 0 & 0 & 0 & 1 \end{pmatrix} \begin{pmatrix} a_{11} & ca_{12}+sa_{13} & ca_{13}-sa_{12} & a_{14} \\ a_{12} & ca_{22}+sa_{23} & ca_{23}-sa_{22} & a_{24} \\ a_{13} & ca_{23}+sa_{33} & ca_{33}-sa_{23} & a_{34} \\ a_{14} & ca_{24}+sa_{34} & ca_{34}-sa_{24} & a_{44} \end{pmatrix}$$

[This has left columns one and four unchanged since we are rotating in the (2, 3)-plane.]

$$= \begin{pmatrix} a_{11} & ca_{12}+sa_{13} & ca_{13}-sa_{12} & a_{14} \\ ca_{12}+sa_{13} & c^2a_{22}+2csa_{23}+s^2a_{33} & (c^2-s^2)a_{23}+cs(a_{33}-a_{22}) & ca_{24}+sa_{34} \\ ca_{13}-sa_{12} & (c^2-s^2)a_{23}+cs(a_{33}-a_{22}) & s^2a_{22}-2csa_{23}+c^2a_{33} & ca_{34}-sa_{24} \\ a_{14} & ca_{24}+sa_{34} & ca_{34}-sa_{24} & a_{44} \end{pmatrix}$$

[This has left rows one and four of the previous matrix unchanged.]

This is, of course, still a symmetric matrix. We should notice that by rotating in the (2, 3)-plane we have not altered those elements of **A** that lie in the intersections of the first and fourth rows and columns, as we would expect. We wish to put

$$ca_{13} - sa_{12} = 0$$

which gives

$$\frac{a_{13}}{a_{12}} = \frac{s}{c} = \tan\theta$$

This means that

$$c = a_{12}(a_{12}^2 + a_{13}^2)^{-\frac{1}{2}} \quad \text{and} \quad s = a_{13}(a_{12}^2 + a_{13}^2)^{-\frac{1}{2}}$$

In order to reduce the element now in the (1, 4) position we rotate in the (2, 4)-plane. From the comments made earlier it is clear that this will not affect the zeros introduced into the (1, 3) and (3, 1) positions.

Lastly we rotate in the (3, 4)-plane in order to reduce to zero the element in the (2, 4) position. It is not obvious that this does not affect the zeros in the (1, 3) and (1, 4) positions, so we shall now demonstrate that this is in fact the case. We have

$$\mathbf{A}_3 = \mathbf{Y}_3^{-1}\mathbf{A}_2\mathbf{Y}_3 = \mathbf{Y}_3^T\mathbf{A}_2\mathbf{Y}_3$$

$$= \begin{pmatrix} 1 & 0 & 0 & 0 \\ 0 & 1 & 0 & 0 \\ 0 & 0 & c & s \\ 0 & 0 & -s & c \end{pmatrix} \begin{pmatrix} b_{11} & b_{12} & 0 & 0 \\ b_{12} & b_{22} & b_{23} & b_{24} \\ 0 & b_{23} & b_{33} & b_{34} \\ 0 & b_{24} & b_{34} & b_{44} \end{pmatrix} \begin{pmatrix} 1 & 0 & 0 & 0 \\ 0 & 1 & 0 & 0 \\ 0 & 0 & c & -s \\ 0 & 0 & s & c \end{pmatrix}$$

$$= \begin{pmatrix} 1 & 0 & 0 & 0 \\ 0 & 1 & 0 & 0 \\ 0 & 0 & c & s \\ 0 & 0 & -s & c \end{pmatrix} \begin{pmatrix} b_{11} & b_{12} & 0 & 0 \\ b_{12} & b_{22} & cb_{23}+sb_{24} & cb_{24}-sb_{23} \\ 0 & b_{23} & cb_{33}+sb_{34} & cb_{34}-sb_{33} \\ 0 & b_{24} & cb_{34}+sb_{44} & cb_{44}-sb_{34} \end{pmatrix}$$

$$= \begin{pmatrix} b_{11} & b_{12} & 0 & 0 \\ b_{12} & b_{22} & cb_{23}+sb_{24} & cb_{24}-sb_{23} \\ 0 & cb_{23}+sb_{24} & c^2 b_{33}+2csb_{34}+s^2 b_{44} & (c^2-s^2)\,b_{34}+cs(b_{44}-b_{33}) \\ 0 & cb_{24}-sb_{23} & (c^2-s^2)\,b_{34}+cs(b_{44}-b_{33}) & c^2 b_{44}-2csb_{34}+s^2 b_{23} \end{pmatrix}$$

If we choose $b_{24}/b_{23} = \tan\theta$ we obtain the required result.

In general we rotate in sequence as follows:

(2, 3)-plane; (2, 4)-plane; (2, 5)-plane; ...; (2, $n-1$)-plane; (2, n)-plane

(3, 4)-plane; (3, 5)-plane; ...; (3, $n-1$)-plane; (3, n)-plane

(4, 5)-plane; ...; (4, $n-1$)-plane; (4, n)-plane

⋮ ⋮

($n-2$, $n-1$)-plane; ($n-2$, n)-plane

($n-1$, n)-plane

Example 6.1

$$\mathbf{A} = \begin{pmatrix} 0 & 3 & 4 \\ 3 & 1 & -1 \\ 4 & -1 & 0 \end{pmatrix}$$

$$c_1 = 3(3^2+4^2)^{-\frac{1}{2}} = \tfrac{3}{5}; \quad s_1 = 4(3^2+4^2)^{-\frac{1}{2}} = \tfrac{4}{5}$$

$\mathbf{A}_1 = \mathbf{Y}_1^T \mathbf{A} \mathbf{Y}_1$

$$= \begin{pmatrix} 1 & 0 & 0 \\ 0 & \frac{3}{5} & \frac{4}{5} \\ 0 & -\frac{4}{5} & \frac{3}{5} \end{pmatrix} \begin{pmatrix} 0 & 3 & 4 \\ 3 & 1 & -1 \\ 4 & -1 & 0 \end{pmatrix} \begin{pmatrix} 1 & 0 & 0 \\ 0 & \frac{3}{5} & -\frac{4}{5} \\ 0 & \frac{4}{5} & \frac{3}{5} \end{pmatrix}$$

$$= \begin{pmatrix} 0 & 3 & 4 \\ 5 & -\frac{1}{5} & -\frac{3}{5} \\ 0 & -\frac{7}{5} & \frac{4}{5} \end{pmatrix} \begin{pmatrix} 1 & 0 & 0 \\ 0 & \frac{3}{5} & -\frac{4}{5} \\ 0 & \frac{4}{5} & \frac{3}{5} \end{pmatrix} = \begin{pmatrix} 0 & 5 & 0 \\ 5 & -\frac{3}{5} & -\frac{1}{5} \\ 0 & -\frac{1}{5} & \frac{8}{5} \end{pmatrix}$$

which is the required tridiagonal form. From Gerschgorin's theorem we see that all the roots lie in the interval $(-6, 6)$. Now

$$f_0(\lambda) = 1$$

$$f_1(\lambda) = -\lambda$$

$$f_2(\lambda) = -(\tfrac{3}{5}+\lambda)f_1(\lambda) - 25$$

$$f_3(\lambda) = (\tfrac{8}{5}-\lambda)f_2(\lambda) - \tfrac{1}{25}f_1(\lambda)$$

x	f_0	f_1	f_2	f_3	$S(x)$	Comment
0	+1	+0	−25	−	1	1 root in $(-6, 0)$, 2 in $(0, 6)$
3	+1	−3	$-\frac{71}{5}$	+	2	1 root in $(0, 3)$, 1 in $(3, 6)$

We have now already established that there is one root in each of the intervals $(-6, 0)$, $(0, 3)$, $(3, 6)$ and we may use any method we choose to converge to particular roots. The actual latent roots are, correct to four decimal places,

$$\lambda_1 = 4{\cdot}7150, \quad \lambda_2 = 1{\cdot}5970, \quad \lambda_3 = -5{\cdot}3121$$

Example 6.2

$$\mathbf{A} = \begin{pmatrix} 7 & 1 & -2 & 1 & 5 \\ 1 & 5 & 0 & 3 & 1 \\ -2 & 0 & 8 & 0 & -2 \\ 1 & 3 & 0 & 5 & 1 \\ 5 & 1 & -2 & 1 & 7 \end{pmatrix}$$

$$c_1 = 1(1^2+2^2)^{-\frac{1}{2}} = \frac{1}{\sqrt{5}}; \quad s_1 = -2(1^2+2^2)^{-\frac{1}{2}} = -\frac{2}{\sqrt{5}}$$

$\mathbf{A}_1 = \mathbf{Y}_1^T \mathbf{A} \mathbf{Y}_1$

$$= \begin{pmatrix} 1 & 0 & 0 & 0 & 0 \\ 0 & \frac{1}{\sqrt{5}} & -\frac{2}{\sqrt{5}} & 0 & 0 \\ 0 & \frac{2}{\sqrt{5}} & \frac{1}{\sqrt{5}} & 0 & 0 \\ 0 & 0 & 0 & 1 & 0 \\ 0 & 0 & 0 & 0 & 1 \end{pmatrix} \begin{pmatrix} 7 & 1 & -2 & 1 & 5 \\ 1 & 5 & 0 & 3 & 1 \\ -2 & 0 & 8 & 0 & -2 \\ 1 & 3 & 0 & 5 & 1 \\ 5 & 1 & -2 & 1 & 7 \end{pmatrix} \begin{pmatrix} 1 & 0 & 0 & 0 & 0 \\ 0 & \frac{1}{\sqrt{5}} & \frac{2}{\sqrt{5}} & 0 & 0 \\ 0 & -\frac{2}{\sqrt{5}} & \frac{1}{\sqrt{5}} & 0 & 0 \\ 0 & 0 & 0 & 1 & 0 \\ 0 & 0 & 0 & 0 & 1 \end{pmatrix}$$

$$= \begin{pmatrix} 7 & \sqrt{5} & 0 & 1 & 5 \\ \sqrt{5} & \frac{37}{5} & -\frac{6}{5} & \frac{3}{\sqrt{5}} & \sqrt{5} \\ 0 & -\frac{6}{5} & \frac{28}{5} & \frac{6}{\sqrt{5}} & 0 \\ 1 & \frac{3}{\sqrt{5}} & \frac{6}{\sqrt{5}} & 5 & 1 \\ 5 & \sqrt{5} & 0 & 1 & 7 \end{pmatrix}$$

$$c_2 = \sqrt{5}(5+1^2)^{-\frac{1}{2}} = \frac{\sqrt{5}}{\sqrt{6}}; \quad s_2 = 1(5+1^2)^{-\frac{1}{2}} = \frac{1}{\sqrt{6}}$$

$\mathbf{A}_2 = \mathbf{Y}_2^T \mathbf{A}_1 \mathbf{Y}_2$

$$= \begin{pmatrix} 1 & 0 & 0 & 0 & 0 \\ 0 & \frac{\sqrt{5}}{\sqrt{6}} & 0 & \frac{1}{\sqrt{6}} & 0 \\ 0 & 0 & 1 & 0 & 0 \\ 0 & -\frac{1}{\sqrt{6}} & 0 & \frac{\sqrt{5}}{\sqrt{6}} & 0 \\ 0 & 0 & 0 & 0 & 1 \end{pmatrix} \begin{pmatrix} 7 & \sqrt{5} & 0 & 1 & 5 \\ \sqrt{5} & \frac{37}{5} & -\frac{6}{5} & \frac{3}{\sqrt{5}} & \sqrt{5} \\ 0 & -\frac{6}{5} & \frac{28}{5} & \frac{6}{\sqrt{5}} & 0 \\ 1 & \frac{3}{\sqrt{5}} & \frac{6}{\sqrt{5}} & 5 & 1 \\ 5 & \sqrt{5} & 0 & 1 & 7 \end{pmatrix} \begin{pmatrix} 1 & 0 & 0 & 0 & 0 \\ 0 & \frac{\sqrt{5}}{\sqrt{6}} & 0 & -\frac{1}{\sqrt{6}} & 0 \\ 0 & 0 & 1 & 0 & 0 \\ 0 & \frac{1}{\sqrt{6}} & 0 & \frac{\sqrt{5}}{\sqrt{6}} & 0 \\ 0 & 0 & 0 & 0 & 1 \end{pmatrix}$$

$$= \begin{pmatrix} 7 & \sqrt{6} & 0 & 0 & 5 \\ \sqrt{6} & 8 & 0 & 0 & \sqrt{6} \\ 0 & 0 & \frac{28}{5} & \frac{6\sqrt{6}}{5} & 0 \\ 0 & 0 & \frac{6\sqrt{6}}{5} & \frac{22}{5} & 0 \\ 5 & \sqrt{6} & 0 & 0 & 7 \end{pmatrix}$$

$$c_3 = \sqrt{6}(6+5^2)^{-\frac{1}{2}} = \frac{\sqrt{6}}{\sqrt{31}}; \quad s_3 = 5(6+5^2)^{-\frac{1}{2}} = \frac{5}{\sqrt{31}}$$

$\mathbf{A}_3 = \mathbf{Y}_3^T \mathbf{A}_2 \mathbf{Y}_3$

$$= \begin{pmatrix} 1 & 0 & 0 & 0 & 0 \\ 0 & \frac{\sqrt{6}}{\sqrt{31}} & 0 & 0 & \frac{5}{\sqrt{31}} \\ 0 & 0 & 1 & 0 & 0 \\ 0 & 0 & 0 & 1 & 0 \\ 0 & -\frac{5}{\sqrt{31}} & 0 & 0 & \frac{\sqrt{6}}{\sqrt{31}} \end{pmatrix} \begin{pmatrix} 7 & \sqrt{6} & 0 & 0 & 5 \\ \sqrt{6} & 8 & 0 & 0 & \sqrt{6} \\ 0 & 0 & \frac{28}{5} & \frac{6\sqrt{6}}{5} & 0 \\ 0 & 0 & \frac{6\sqrt{6}}{5} & \frac{22}{5} & 0 \\ 5 & \sqrt{6} & 0 & 0 & 7 \end{pmatrix} \begin{pmatrix} 1 & 0 & 0 & 0 & 0 \\ 0 & \frac{\sqrt{6}}{\sqrt{31}} & 0 & 0 & -\frac{5}{\sqrt{31}} \\ 0 & 0 & 1 & 0 & 0 \\ 0 & 0 & 0 & 1 & 0 \\ 0 & \frac{5}{\sqrt{31}} & 0 & 0 & \frac{\sqrt{6}}{\sqrt{31}} \end{pmatrix}$$

$$= \begin{pmatrix} 7 & \sqrt{31} & 0 & 0 & 0 \\ \sqrt{31} & \frac{283}{31} & 0 & 0 & -\frac{24\sqrt{6}}{31} \\ 0 & 0 & \frac{28}{5} & \frac{6\sqrt{6}}{5} & 0 \\ 0 & 0 & \frac{6\sqrt{6}}{5} & \frac{22}{5} & 0 \\ 0 & -\frac{24\sqrt{6}}{31} & 0 & 0 & \frac{182}{31} \end{pmatrix}$$

Note that the (2, 4) position is already zero, so that we proceed to the (2, 5) position next.

$$c_4 = 0[0^2 + (-\tfrac{24\sqrt{6}}{31})^2]^{-\frac{1}{2}} = 0; \quad s_4 = -\tfrac{24\sqrt{6}}{31}[0^2 + (-\tfrac{24\sqrt{6}}{31})^2]^{-\frac{1}{2}} = -1$$

$\mathbf{A}_4 = \mathbf{Y}_4^T \mathbf{A}_3 \mathbf{Y}_4$

$$= \begin{pmatrix} 1 & 0 & 0 & 0 & 0 \\ 0 & 1 & 0 & 0 & 0 \\ 0 & 0 & 0 & 0 & -1 \\ 0 & 0 & 0 & 1 & 0 \\ 0 & 0 & 1 & 0 & 0 \end{pmatrix} \begin{pmatrix} 7 & \sqrt{31} & 0 & 0 & 0 \\ \sqrt{31} & \frac{283}{31} & 0 & 0 & -\frac{24\sqrt{6}}{31} \\ 0 & 0 & \frac{28}{5} & \frac{6\sqrt{6}}{5} & 0 \\ 0 & 0 & \frac{6\sqrt{6}}{5} & \frac{22}{5} & 0 \\ 0 & -\frac{24\sqrt{6}}{31} & 0 & 0 & \frac{182}{31} \end{pmatrix} \begin{pmatrix} 1 & 0 & 0 & 0 & 0 \\ 0 & 1 & 0 & 0 & 0 \\ 0 & 0 & 0 & 0 & 1 \\ 0 & 0 & 0 & 1 & 0 \\ 0 & 0 & -1 & 0 & 0 \end{pmatrix}$$

$$= \begin{pmatrix} 7 & \sqrt{31} & 0 & 0 & 0 \\ \sqrt{31} & \frac{283}{31} & \frac{24\sqrt{6}}{31} & 0 & 0 \\ 0 & \frac{24\sqrt{6}}{31} & \frac{182}{31} & 0 & 0 \\ 0 & 0 & 0 & \frac{22}{5} & \frac{6\sqrt{6}}{5} \\ 0 & 0 & 0 & \frac{6\sqrt{6}}{5} & \frac{28}{5} \end{pmatrix}$$

This is now in tridiagonal form. We cannot use the Sturm theory directly on $\mathbf{A}_4$ because there is a zero superdiagonal element. As in example 5.3 we can partition $\mathbf{A}_4$ and apply the Sturm theory to each of the tridiagonal submatrices. Here we have

$$\mathbf{D}_1 = \begin{pmatrix} 7 & \sqrt{31} & 0 \\ \sqrt{31} & \frac{283}{31} & \frac{24\sqrt{6}}{31} \\ 0 & \frac{24\sqrt{6}}{31} & \frac{182}{31} \end{pmatrix} \quad \text{and} \quad \mathbf{D}_3 = \begin{pmatrix} \frac{22}{5} & \frac{6\sqrt{6}}{6} \\ \frac{6\sqrt{6}}{5} & \frac{28}{5} \end{pmatrix}$$

The roots of $\mathbf{D}_1$ are actually $\lambda_1 = 2$, $\lambda_2 = 6$, $\lambda_3 = 14$ and the roots of $\mathbf{D}_3$ are $\lambda_4 = 2$, $\lambda_5 = 8$.

6.3 Latent Vectors and Givens' Method

If $\mathbf{Y}$ is a latent vector of the tridiagonal matrix we have

$$\mathbf{BY} = \begin{pmatrix} a_1 & b_2 & 0 & \dots & 0 & 0 & 0 \\ b_2 & a_2 & b_3 & \dots & 0 & 0 & 0 \\ \vdots & \vdots & \vdots & & \vdots & \vdots & \vdots \\ 0 & 0 & 0 & \dots & b_{n-1} & a_{n-1} & b_n \\ 0 & 0 & 0 & \dots & 0 & b_n & a_n \end{pmatrix} \begin{pmatrix} y_1 \\ y_2 \\ \vdots \\ y_{n-1} \\ y_n \end{pmatrix} = \lambda \begin{pmatrix} y_1 \\ y_2 \\ \vdots \\ y_{n-1} \\ y_n \end{pmatrix} = \lambda \mathbf{Y}$$

which yield the set of equations

$$\begin{aligned} a_1 y_1 + b_2 y_2 &= \lambda y_1 \\ b_2 y_1 + a_2 y_2 + b_3 y_3 &= \lambda y_2 \\ \ddots \qquad &\quad \vdots \\ b_{n-1} y_{n-2} + a_{n-1} y_{n-1} + b_n y_n &= \lambda y_{n-1} \\ b_n y_{n-1} + a_n y_n &= \lambda y_n \end{aligned}$$

If we arbitrarily put $y_1 = 1$ we can see that the first equation can then be solved for y_2, which then allows the second equation to be solved for y_3, and so on. We shall then be left with the nth equation which we may use as a check.

Example 6.3

Taking the matrix $\mathbf{A}$ of example 6.1 which had latent roots, correct to four decimal places, $\lambda_1 = 4{\cdot}7150$, $\lambda_2 = 1{\cdot}5970$, $\lambda_3 = -5{\cdot}3121$, we have, using $\mathbf{A}_1 \mathbf{Y} = \lambda \mathbf{Y}$,

$$\left.\begin{aligned} 5y_2 \qquad\quad &= \lambda y_1 \\ 5y_1 - \tfrac{3}{5}y_2 - \tfrac{1}{5}y_3 &= \lambda y_2 \\ -\tfrac{1}{5}y_2 + \tfrac{8}{5}y_3 &= \lambda y_3 \end{aligned}\right\} \tag{6.1}$$

When $\lambda = 4{\cdot}715$, putting $y_1 = 1$ we get

$$y_2 = \frac{4{\cdot}715}{5} = 0{\cdot}943$$

and

$$y_3 = 5(5-\tfrac{3}{5}y_2-4{\cdot}715y_2) = -0{\cdot}060225$$

From the third equation we get

$$y_3 = \frac{\frac{1}{5}y_2}{\frac{8}{5}-\lambda} = -0{\cdot}06055$$

and so it would seem that we can only guarantee two-decimal-place accuracy. As it happens the second value of y_3 is correct. Taking

$$\mathbf{Y}_1 = \begin{pmatrix} 1 \\ 0{\cdot}94 \\ -0{\cdot}06 \end{pmatrix}$$

we get

$$\mathbf{X}_1 = \begin{pmatrix} 1 & 0 & 0 \\ 0 & \frac{3}{5} & -\frac{4}{5} \\ 0 & \frac{4}{5} & \frac{3}{5} \end{pmatrix} \begin{pmatrix} 1 \\ 0{\cdot}94 \\ -0{\cdot}06 \end{pmatrix} = \begin{pmatrix} 1 \\ 0{\cdot}612 \\ 0{\cdot}716 \end{pmatrix} \doteqdot \begin{pmatrix} 1 \\ 0{\cdot}61 \\ 0{\cdot}72 \end{pmatrix}$$

Multiplying $\mathbf{X}_1$ by the first row of $\mathbf{A}$ we get

$$(\,0 \quad 3 \quad 4\,) \begin{pmatrix} 1 \\ 0{\cdot}61 \\ 0{\cdot}72 \end{pmatrix} = 4{\cdot}71$$

which agrees as well as can be expected with the latent root.

When $\lambda = 1{\cdot}597$, putting $y_1 = 1$ we get

$$y_2 = 0{\cdot}3194$$

and

$$y_3 = 21{\cdot}4914$$

From the third of equations (6.1) we get

$$y_3 = 21{\cdot}2933$$

and again we have a breakdown in accuracy. This time it is the first value of y_3 which is closest!

This example demonstrates well the difficulty of the latent vector problem in tridiagonal matrices. In other respects the matrix is quite well conditioned because the latent roots are well separated and the matrix is symmetrical, and at first sight we would certainly not expect this loss of accuracy in the latent vectors. Wilkinson† shows that this will often be so. He also discusses how we may improve the accuracy and suggests the method of inverse iteration, which is discussed in § 9.4.

† See reference 7, pp. 315–323.

6.4 Number of Calculations Required by the Givens Method

In the Givens method we have to calculate a square root at each stage which makes it a little difficult to compare with other methods. For the sake of comparison we shall assume that a square root is equivalent to two ordinary calculations. Then to find $\cos\theta$ at each stage will require five calculations, and $\sin\theta$ will require only one. So in the full $\frac{1}{2}(n-1)(n-2)$ transformations we shall perform

$$3(n-1)(n-2) \text{ calculations}$$

In reducing the a_{13} position to zero we perform $4n-2$ multiplications in forming $\mathbf{AY}_1$, and a further six multiplications only in forming $\mathbf{Y}_1^T\mathbf{AY}_1$, providing that we take full advantage of symmetry. We require the same number of calculations for each position in the first row, so that this requires

$$(4n-2)(n-2)+6(n-2) \text{ calculations}$$

The second row will require

$$[4(n-1)-2](n-3)+6(n-3) \text{ calculations}$$

In all the number of calculations required are given by

$$\begin{aligned} \mathbf{S}_n &= 3(n-1)(n-2)+\{(4n-2)(n-2)+6(n-2)+[4(n-1)-2](n-3) \\ &\qquad +6(n-3)+[4(n-2)-2](n-4)+6(n-4)+\ldots+(4\cdot3-2)+6\} \\ &= 3(n-1)(n-2)+6[(n-2)+(n-3)+\ldots+1] \\ &\qquad +(4n-2)(n-2)+(4n-6)(n-3)+(4n-10)(n-4)+\ldots+10 \\ &= 3(n-1)(n-2)+3(n-1)(n-2)+\tfrac{1}{3}(n-2)(n-1)(3+4n) \\ &= \tfrac{1}{3}(n-2)(n-1)(21+4n) \end{aligned}$$

(See table 6.1.)

Table 6.1

n	$\frac{1}{3}(n-2)(n-1)(21+4n)$
3	22
4	74
5	164
6	300
7	490
8	742
9	1064
10	1464
20	11 514
50	173 264
100	1 361 514

The number of calculations required to find the latent vectors will of course depend on the method used to solve the set of equations obtained when performing inverse iteration and on the number of iterations needed.

6.5 Further Comments on Givens' Method

A comparison of Tables 6.1 and 3.1 shows, somewhat surprisingly, that Danilevsky's method, which takes no account of symmetry, requires fewer calculations than that of Givens. Against this we must, of course, take into account that the location of the latent roots is more convenient when we have a tridiagonal matrix, especially if we only require particular latent roots such as those in a given interval. Also, Givens' method is a stable one for determining latent roots.†

Providing that we are able to find square roots accurately, the Givens method is quite convenient for hand calculations because it is easy to remember and simple to perform. If we are using a computer we would almost certainly use Householder's method instead. This is the next method to be discussed.

6.6 Exercises

6.1. Show that the matrix

$$\mathbf{A} = \begin{pmatrix} 1 & 0 & \dots & 0 & \dots & 0 & \dots & 0 & 0 \\ 0 & 1 & \dots & 0 & \dots & 0 & \dots & 0 & 0 \\ \vdots & \vdots & & \vdots & & \vdots & & \vdots & \vdots \\ 0 & 0 & \dots & \cos\theta & \dots & -\sin\theta & \dots & 0 & 0 \\ \vdots & \vdots & & \vdots & & \vdots & & \vdots & \vdots \\ 0 & 0 & \dots & \sin\theta & \dots & \cos\theta & \dots & 0 & 0 \\ \vdots & \vdots & & \vdots & & \vdots & & \vdots & \vdots \\ 0 & 0 & & 0 & & 0 & \dots & 1 & 0 \\ 0 & 0 & \dots & 0 & \dots & 0 & \dots & 0 & 1 \end{pmatrix} \begin{matrix} \\ \\ \\ \text{row } p \\ \\ \text{row } q \\ \\ \\ \\ \end{matrix}$$

$$\text{column } p \qquad \text{column } q$$

is orthogonal.

6.2. Use the method of Givens to find the latent roots and vectors of the following matrices:

$$\text{(i)}\quad \mathbf{A} = \begin{pmatrix} 6 & 3 & 4 \\ 3 & 6 & 4 \\ 4 & 4 & 5 \end{pmatrix}; \qquad \text{(ii)}\quad \mathbf{A} = \begin{pmatrix} 1 & 8 & -6 \\ 8 & -11 & 12 \\ -6 & 12 & -4 \end{pmatrix}$$

† See reference 7, pp. 286–290.

6.3. Use the method of Givens to transform the matrix

$$\mathbf{A} = \begin{pmatrix} 2 & 4 & 3 & 0 \\ 4 & -2 & -3 & 2 \\ 3 & -3 & 6 & 4 \\ 0 & 2 & 4 & 3 \end{pmatrix}$$

into a tridiagonal matrix **B**. Use Sturm's theorem to locate the latent roots of **B** to the nearest integer. Find the smallest roots correct to two decimal places. Find the latent vectors of **B** corresponding to these latent roots and hence the corresponding latent vectors of **A**.

6.4. Show that the sum of squares of the elements of $\mathbf{A}_{r+1}$ is the same as that of $\mathbf{A}_r$ for the Givens transformation. That is, the Euclidean norm, $\|\mathbf{A}\|_E$, is preserved in transforming **A** to tridiagonal form. This is one of the reasons for the stability of the method, because it means that the size of the elements is bounded at each stage, which certainly is not true for Danilevsky's method.

Chapter 7

THE METHOD OF HOUSEHOLDER

The method of Householder, as with Givens', reduces a symmetric matrix to tridiagonal form by means of a series of orthogonal similarity transformations. Whereas Givens reduces a single element to zero at each stage, Householder introduces the required zeros into a whole row at each stage.

7.1 A Symmetric Orthogonal Matrix

Consider the matrix $\mathbf{P}$ given by

$$\mathbf{P} = \mathbf{I} - 2\mathbf{Y}\mathbf{Y}^T$$

where $\mathbf{Y}$ is a column vector such that $\mathbf{Y}^T\mathbf{Y} = 1$. Clearly $\mathbf{P}$ is symmetric so that

$$\begin{aligned}\mathbf{P}\mathbf{P}^T &= \mathbf{P}^2 = (\mathbf{I} - 2\mathbf{Y}\mathbf{Y}^T)^2 \\ &= \mathbf{I} - 2\mathbf{Y}\mathbf{Y}^T - 2\mathbf{Y}\mathbf{Y}^T + 4\mathbf{Y}\mathbf{Y}^T\mathbf{Y}\mathbf{Y}^T \\ &= \mathbf{I} - 4\mathbf{Y}\mathbf{Y}^T + 4\mathbf{Y}\mathbf{Y}^T \\ &= \mathbf{I}\end{aligned}$$

Hence $\mathbf{P}$ is an orthogonal matrix with the special property that

$$\mathbf{P}^{-1} = \mathbf{P}^T = \mathbf{P}$$

It is an orthogonal matrix of this form that is used at each stage of the Householder reduction.

7.2 The Method of Householder

We take as representative the four-by-four matrix given by

$$\mathbf{A} = \begin{pmatrix} a_{11} & a_{12} & a_{13} & a_{14} \\ a_{12} & a_{22} & a_{23} & a_{24} \\ a_{13} & a_{23} & a_{33} & a_{34} \\ a_{14} & a_{24} & a_{34} & a_{44} \end{pmatrix}$$

We wish to reduce a_{13} and a_{14} to zero. The method of Givens suggests that we should be concerned with the $(2, 3)$ and $(2, 4)$-planes. For this reason we take

$$\mathbf{Y}_1^T = (\,0 \quad y_2 \quad y_3 \quad y_4\,)$$

where $y_2^2+y_3^2+y_4^2 = 1$, so that

$$\mathbf{Y}_1\mathbf{Y}_1^T = \begin{pmatrix} 0 \\ y_2 \\ y_3 \\ y_4 \end{pmatrix} (0 \quad y_2 \quad y_3 \quad y_4) = \begin{pmatrix} 0 & 0 & 0 & 0 \\ 0 & y_2^2 & y_2 y_3 & y_2 y_4 \\ 0 & y_2 y_3 & y_3^2 & y_3 y_4 \\ 0 & y_2 y_4 & y_3 y_4 & y_4^2 \end{pmatrix}$$

and

$$\mathbf{P}_1 = \mathbf{I}-2\mathbf{Y}_1\mathbf{Y}_1^T = \begin{pmatrix} 1 & 0 & 0 & 0 \\ 0 & 1-2y_2^2 & -2y_2 y_3 & -2y_2 y_4 \\ 0 & -2y_2 y_3 & 1-2y_3^2 & -2y_3 y_4 \\ 0 & -2y_2 y_4 & -2y_3 y_4 & 1-2y_4^2 \end{pmatrix}$$

Forming $\mathbf{P}_1^T\mathbf{A}\mathbf{P}_1 = \mathbf{P}_1\mathbf{A}\mathbf{P}_1$ and denoting its elements by b_{ij} we find that

$$b_{12} = a_{12}-2y_2(a_{12}y_2+a_{13}y_3+a_{14}y_4) = a_{12}-2y_2 z \tag{7.1}$$

$$b_{13} = a_{13}-2y_3(a_{12}y_2+a_{13}y_3+a_{14}y_4) = a_{13}-2y_3 z \tag{7.2}$$

$$b_{14} = a_{14}-2y_4(a_{12}y_2+a_{13}y_3+a_{14}y_4) = a_{14}-2y_4 z \tag{7.3}$$

where $z = a_{12}y_2+a_{13}y_3+a_{14}y_4$. Squaring each of the equations (7.1), (7.2), (7.3) and adding we get

$$\begin{aligned} b_{12}^2+b_{13}^2+b_{14}^2 &= (a_{12}-2y_2 z)^2+(a_{13}-2y_3 z)^2+(a_{14}-2y_4 z)^2 \\ &= a_{12}^2+a_{13}^2+a_{14}^2-4z(a_{12}y_2+a_{13}y_3+a_{14}y_4)+4z^2(y_2^2+y_3^2+y_4^2) \\ &= a_{12}^2+a_{13}^2+a_{14}^2-4z^2+4z^2 \\ &= a_{12}^2+a_{13}^2+a_{14}^2 \end{aligned} \tag{7.4}$$

We require that $b_{13} = b_{14} = 0$. Hence from equations (7.2) and (7.3) we get

$$a_{13}-2y_3 z = a_{14}-2y_4 z = 0 \tag{7.5}$$

Also, from equations (7.1) and (7.4) we have

$$a_{12}^2+a_{13}^2+a_{14}^2 = 0^2+0^2+(a_{12}-2y_2 z)^2 \tag{7.6}$$

so that

$$a_{12}-2y_2 z = \pm(a_{12}^2+a_{13}^2+a_{14}^2)^{\frac{1}{2}}$$

and

$$a_{12}y_2-2y_2^2 z = \pm y_2(a_{12}^2+a_{13}^2+a_{14}^2)^{\frac{1}{2}} \tag{7.7}$$

Multiplying the two equations of (7.5) by y_3 and y_4 respectively and adding them to (7.7) we find

$$(a_{12}y_2+a_{13}y_3+a_{14}y_4)-2z(y_2^2+y_3^2+y_4^2) = \pm y_2(a_{12}^2+a_{13}^2+a_{14}^2)^{\frac{1}{2}}$$

so that

$$z-2z = -z = \pm y_2(a_{12}^2+a_{13}^2+a_{14}^2)^{\frac{1}{2}}$$

or

$$z = \pm y_2(a_{12}^2+a_{13}^2+a_{14}^2)^{\frac{1}{2}}$$

Substituting for z in (7.5) and (7.6) we finally find

$$y_2^2 = \frac{1}{2}\left[1 \pm \frac{a_{12}}{(a_{12}^2+a_{13}^2+a_{14}^2)^{\frac{1}{2}}}\right] \tag{7.8}$$

$$y_3 = \pm\frac{a_{13}}{2y_2(a_{12}^2+a_{13}^2+a_{14}^2)^{\frac{1}{2}}}$$

$$y_4 = \pm\frac{a_{14}}{2y_2(a_{12}^2+a_{13}^2+a_{14}^2)^{\frac{1}{2}}}$$

It is usual to select the sign in equation (7.8) so that y_2^2 is as large as possible, and in particular so that $y_2 \neq 0$.

In order to reduce the b_{24} position of $\mathbf{A}_1$ to zero we can use either a transformation of this type or the slightly simpler Givens transformation.

It is hoped that the extension to the general case is now clear.

Example 7.1

$$\mathbf{A} = \begin{pmatrix} -1 & 1 & 2 & 2 \\ 1 & 3 & -4 & -4 \\ 2 & -4 & -3 & -3 \\ 2 & -4 & -3 & -3 \end{pmatrix}$$

$$y_2^2 = \frac{1}{2}\left[1+\frac{1}{(1^2+2^2+2^2)^{\frac{1}{2}}}\right] = \frac{1}{2}\left(1+\frac{1}{3}\right) = \frac{2}{3}$$

so that

$$y_2 = \sqrt{\left(\frac{2}{3}\right)} = \frac{\sqrt{6}}{3}$$

$$y^3 = \frac{2}{2.\frac{\sqrt{6}}{3}.3} = \frac{1}{\sqrt{6}} = \frac{\sqrt{6}}{6}$$

$$y_4 = \frac{\sqrt{6}}{6}$$

Hence

$\mathbf{A}_1 = \mathbf{P}_1 \mathbf{A} \mathbf{P}_1$

$$= \begin{pmatrix} 1 & 0 & 0 & 0 \\ 0 & -\frac{1}{3} & -\frac{2}{3} & -\frac{2}{3} \\ 0 & -\frac{2}{3} & \frac{2}{3} & -\frac{1}{3} \\ 0 & -\frac{2}{3} & -\frac{1}{3} & \frac{2}{3} \end{pmatrix} \begin{pmatrix} -1 & 1 & 2 & 2 \\ 1 & 3 & -4 & -4 \\ 2 & -4 & -3 & -3 \\ 2 & -4 & -3 & -3 \end{pmatrix} \begin{pmatrix} 1 & 0 & 0 & 0 \\ 0 & -\frac{1}{3} & -\frac{2}{3} & -\frac{2}{3} \\ 0 & -\frac{2}{3} & \frac{2}{3} & -\frac{1}{3} \\ 0 & -\frac{2}{3} & -\frac{1}{3} & \frac{2}{3} \end{pmatrix}$$

$$= \begin{pmatrix} 1 & 0 & 0 & 0 \\ 0 & -\frac{1}{3} & -\frac{2}{3} & -\frac{2}{3} \\ 0 & -\frac{2}{3} & \frac{2}{3} & -\frac{1}{3} \\ 0 & -\frac{2}{3} & -\frac{1}{3} & \frac{2}{3} \end{pmatrix} \begin{pmatrix} -1 & -3 & 0 & 0 \\ 1 & \frac{13}{3} & -\frac{10}{3} & -\frac{10}{3} \\ 2 & \frac{16}{3} & \frac{5}{3} & \frac{5}{3} \\ 2 & \frac{16}{3} & \frac{5}{3} & \frac{5}{3} \end{pmatrix}$$

$$= \begin{pmatrix} -1 & -3 & 0 & 0 \\ -3 & -\frac{77}{9} & -\frac{10}{9} & -\frac{10}{9} \\ 0 & -\frac{10}{9} & \frac{25}{9} & \frac{25}{9} \\ 0 & -\frac{10}{9} & \frac{25}{9} & \frac{25}{9} \end{pmatrix}$$

Now

$$y_3^2 = \frac{1}{2}\left[1 - \frac{-\frac{10}{9}}{[(-\frac{10}{9})^2 + (-\frac{10}{9})^2]^{\frac{1}{2}}}\right] = \frac{1}{2}\left(1 + \frac{1}{\sqrt{2}}\right)$$

$$= \frac{1+\sqrt{2}}{2\sqrt{2}} = \frac{\sqrt{2}+2}{4}$$

so that

$$y_3 = \frac{(\sqrt{2}+2)^{\frac{1}{2}}}{2}$$

$$y_4 = \frac{\frac{10}{9}}{2y_3 . \frac{10\sqrt{2}}{9}} = \frac{1}{\sqrt{2}(\sqrt{2}+2)^{\frac{1}{2}}} = \frac{(\sqrt{2}+2)^{\frac{1}{2}}}{\sqrt{2}(\sqrt{2}+2)}$$

Hence

$$\mathbf{A}_2 = \mathbf{P}_2\mathbf{A}_1\mathbf{P}_2$$

$$= \begin{pmatrix} 1 & 0 & 0 & 0 \\ 0 & 1 & 0 & 0 \\ 0 & 0 & -\frac{\sqrt{2}}{2} & -\frac{\sqrt{2}}{2} \\ 0 & 0 & -\frac{\sqrt{2}}{2} & \frac{\sqrt{2}}{2} \end{pmatrix} \begin{pmatrix} -1 & -3 & 0 & 0 \\ -3 & -\frac{77}{9} & -\frac{10}{9} & -\frac{10}{9} \\ 0 & -\frac{10}{9} & \frac{25}{9} & \frac{25}{9} \\ 0 & -\frac{10}{9} & \frac{25}{9} & \frac{25}{9} \end{pmatrix} \begin{pmatrix} 1 & 0 & 0 & 0 \\ 0 & 1 & 0 & 0 \\ 0 & 0 & -\frac{\sqrt{2}}{2} & -\frac{\sqrt{2}}{2} \\ 0 & 0 & -\frac{\sqrt{2}}{2} & \frac{\sqrt{2}}{2} \end{pmatrix}$$

$$= \begin{pmatrix} 1 & 0 & 0 & 0 \\ 0 & 1 & 0 & 0 \\ 0 & 0 & -\frac{\sqrt{2}}{2} & -\frac{\sqrt{2}}{2} \\ 0 & 0 & -\frac{\sqrt{2}}{2} & \frac{\sqrt{2}}{2} \end{pmatrix} \begin{pmatrix} -1 & -3 & 0 & 0 \\ -3 & -\frac{77}{9} & \frac{10\sqrt{2}}{9} & 0 \\ 0 & -\frac{10}{9} & -\frac{25\sqrt{2}}{9} & 0 \\ 0 & -\frac{10}{9} & -\frac{25\sqrt{2}}{9} & 0 \end{pmatrix}$$

$$= \begin{pmatrix} -1 & -3 & 0 & 0 \\ -3 & -\frac{77}{9} & \frac{10\sqrt{2}}{9} & 0 \\ 0 & \frac{10\sqrt{2}}{9} & \frac{50}{9} & 0 \\ 0 & 0 & 0 & 0 \end{pmatrix}$$

which is the required tridiagonal form, and we may find the latent roots in the usual way. The roots are

$$\lambda_1 = \lambda_2 = 0; \quad \lambda_3, \lambda_4 = 2(-1 \pm \sqrt{15})$$

We note again the multiple root causing the zero superdiagonal elements.

7.3 Reducing the Number of Calculations

To gain the most benefit from Householder's method we need to give careful consideration to its execution.

Let the matrix $\mathbf{A}_{r-1}$ have elements a_{ij}, and put

$$b = (a^2_{r,r+1} + a^2_{r,r+2} + \ldots + a^2_{rn})^{\frac{1}{2}}$$

and

$$d = b + |a_{r,r+1}|$$

Then we get for the components of $\mathbf{Y}_r$

$$y_{r+1} = \left(\frac{d}{2b}\right)^{\frac{1}{2}}$$

and

$$y_j = \pm\frac{a_{rj}}{2by_{r+1}} = \pm\frac{a_{rj}}{(2bd)^{\frac{1}{2}}}, \quad j = r+2, r+3, \ldots, n$$

the sign chosen to be that of $a_{r,r+1}$. If we let $\mathbf{Z}_r = (2bd)^{\frac{1}{2}}\mathbf{Y}_r$, then the components of $\mathbf{Z}_r$ are

$$z_{r+1} = d$$

and

$$z_j = \pm a_{rj}, \quad j = r+2, r+3, \ldots, n$$

Further, if we put

$$\mathbf{W}_r = \frac{1}{bd}\mathbf{A}_{r-1}\mathbf{Z}_r$$

and

$$\mathbf{V}_r = \mathbf{W}_r - \frac{1}{2bd}\mathbf{Z}_r\mathbf{Z}_r^T\mathbf{W}_r$$

we get

$$\begin{aligned}
\mathbf{A}_r &= \mathbf{P}_r\mathbf{A}_{r-1}\mathbf{P}_r \\
&= (\mathbf{I}-2\mathbf{Y}_r\mathbf{Y}_r^T)\mathbf{A}_{r-1}(\mathbf{I}-2\mathbf{Y}_r\mathbf{Y}_r^T) \\
&= \mathbf{A}_{r-1} - 2\mathbf{A}_{r-1}\mathbf{Y}_r\mathbf{Y}_r^T - 2\mathbf{Y}_r\mathbf{Y}_r^T\mathbf{A}_{r-1} + 4\mathbf{Y}_r\mathbf{Y}_r^T\mathbf{A}_{r-1}\mathbf{Y}_r\mathbf{Y}_r^T \\
&= \mathbf{A}_{r-1} - 2(\mathbf{A}_{r-1}\mathbf{Y}_r\mathbf{Y}_r^T - \mathbf{Y}_r\mathbf{Y}_r^T\mathbf{A}_{r-1}\mathbf{Y}_r\mathbf{Y}_r^T) - 2(\mathbf{Y}_r\mathbf{Y}_r^T\mathbf{A}_{r-1} - \mathbf{Y}_r\mathbf{Y}_r^T\mathbf{A}_{r-1}\mathbf{Y}_r\mathbf{Y}_r^T) \\
&= \mathbf{A}_{r-1} - \frac{1}{bd}\left(\mathbf{A}_{r-1}\mathbf{Z}_r\mathbf{Z}_r^T - \frac{1}{2bd}\mathbf{Z}_r\mathbf{Z}_r^T\mathbf{A}_{r-1}\mathbf{Z}_r\mathbf{Z}_r^T\right) \\
&\qquad - \frac{1}{bd}\left(\mathbf{Z}_r\mathbf{Z}_r^T\mathbf{A}_{r-1} - \frac{1}{2bd}\mathbf{Z}_r\mathbf{Z}_r^T\mathbf{A}_{r-1}\mathbf{Z}_r\mathbf{Z}_r^T\right) \\
&= \mathbf{A}_{r-1} - \left(\mathbf{W}_r\mathbf{Z}_r^T - \frac{1}{2bd}\mathbf{Z}_r\mathbf{Z}_r^T\mathbf{W}_r\mathbf{Z}_r^T\right) - \left(\mathbf{Z}_r\mathbf{W}_r^T - \frac{1}{2bd}\mathbf{Z}_r\mathbf{W}_r^T\mathbf{Z}_r\mathbf{Z}_r^T\right) \\
&= \mathbf{A}_{r-1} - \mathbf{V}_r\mathbf{Z}_r^T - \mathbf{Z}_r\mathbf{V}_r^T \\
&= \mathbf{A}_{r-1} - (\mathbf{X}_r + \mathbf{X}_r^T)
\end{aligned}$$

where

$$\mathbf{X}_r = \mathbf{V}_r\mathbf{Z}_r^T$$

So we compute in sequence $\mathbf{W}_r$, $\mathbf{V}_r$ and $\mathbf{X}_r$. Since $\mathbf{A}_r$ is symmetric we do not need the elements of $\mathbf{X}_r$ below its leading diagonal. Also, we only require the last $(n-r)^2$ elements of $\mathbf{A}_r$.

7.4 Number of Calculations Required by Householder's Method

We first find the number of calculations required to obtain $\mathbf{A}_r$, and again take a square root as being two calculations.

To find $\mathbf{A}_{r-1}\mathbf{Z}_r$ requires $(n-r)^2$ calculations, since we require only the last $(n-r)$ elements. So, to find $\mathbf{W}_r$ takes $(n-r)^2 + (n-r+3) + (n-r)$ calculations.

Bearing in mind that $\mathbf{Z}_r^T \mathbf{W}_r$ contains only a single element, to find $\mathbf{V}_r$ takes a further $2(n-r+1)+1$ calculations. Lastly, to obtain the required elements of $\mathbf{X}_r$ takes $(n-r)^2$ calculations. So, in all, $\mathbf{A}_r$ requires

$$(n-r)^2+(n-r+3)+(n-r)+2(n-r+1)+1+(n-r)^2$$
$$= 2(n-r)^2+4(n-r)+6 \quad \text{calculations}$$

In all, the method of Householder requires S calculations, where

$$\begin{aligned} S &= \sum_{r=1}^{n-2} [2(n-r)^2+4(n-r)+6] \\ &= [\tfrac{1}{3}n(n-1)(2n-1)-2]+[2n(n-1)-4]+6(n-2) \\ &= \tfrac{1}{3}(2n^3-3n^2+n-6+6n^2-6n-12+18n-36) \\ &= \tfrac{1}{3}(2n^3+3n^2+13n-54) \\ &= \tfrac{1}{3}(n-2)(2n^2+7n+27) \end{aligned}$$

(See Table 7.1.)

TABLE 7.1

n	$\frac{1}{3}(n-2)(2n^2+7n+27)$
3	22
4	58
5	112
6	188
7	290
8	422
9	588
10	792
20	5802
50	86032
100	677082

7.5 Further Comments on Householder's Method

Householder's method requires about half the number of calculations that Givens' does, and is also a substantial improvement on Danilevsky's method. By comparison with Givens' method, we can see that, for use on a computer, we prefer the method of Householder, which is faster and requires less storage. As a hand method, Householder's is rather complex, and in this case we may still prefer Givens' method. We also note that, if any element is already zero, we do not have to perform that particular transformation with Givens' method, but we can only miss out a Householder transformation if all the required elements in a row are zero.

As with Given's method, Householder's method gives a stable reduction to tridiagonal form.

7.6 Exercises

7.1. Use the method of Householder to reduce the following matrices to tridiagonal form:

$$\text{(i)}\quad \mathbf{A} = \begin{pmatrix} 1 & 7 & 24 \\ 7 & 5 & 12 \\ 24 & 12 & -2 \end{pmatrix} \qquad \text{(ii)}\quad \mathbf{A} = \begin{pmatrix} 5 & 7 & 4 & -4 \\ 7 & 1 & 9 & -4 \\ 4 & 9 & 4 & 2 \\ -4 & -4 & 2 & -3 \end{pmatrix}$$

7.2. Show that the property of exercise 6.4 applies also to the Householder transformation.

Chapter 8

THE METHOD OF LANCZOS

The method of Lanczos also reduces a matrix to tridiagonal form by means of a similarity transformation. It finds the matrix producing the transformation, not as a finite sequence of transformations, but by constructing the sequence of column vectors that constitute this matrix.

8.1 The Method of Lanczos for Symmetric Matrices

Lanczos' method reduces the matrix to the tridiagonal matrix $\mathbf{C}$, which is given by

$$\mathbf{C} = \begin{pmatrix} a_1 & b_2 & 0 & 0 & \dots & 0 & 0 & 0 \\ 1 & a_2 & b_3 & 0 & \dots & 0 & 0 & 0 \\ 0 & 1 & a_3 & b_4 & \dots & 0 & 0 & 0 \\ \vdots & \vdots & \vdots & \vdots & & \vdots & \vdots & \vdots \\ 0 & 0 & 0 & 0 & \dots & 0 & 1 & a_n \end{pmatrix}$$

by finding $\mathbf{Y}$ such that $\mathbf{C} = \mathbf{Y}^{-1}\mathbf{A}\mathbf{Y}$. The method constructs the matrix $\mathbf{Y}$ as a sequence of column vectors $\mathbf{Y}_1, \mathbf{Y}_2, \dots, \mathbf{Y}_n$, where each $\mathbf{Y}_i$ is orthogonal to each of the previous vectors in the sequence, and $\mathbf{Y}_{i+1}$ is given by the recurrence relation

$$\mathbf{Y}_{i+1} = \mathbf{A}\mathbf{Y}_i - a_i\mathbf{Y}_i - b_i\mathbf{Y}_{i-1} \tag{8.1}$$

where $\mathbf{Y}_0 = 0$ and $\mathbf{Y}_1$ is arbitrary, and a_i, b_i are chosen so that $\mathbf{Y}_{i+1}$ is orthogonal to $\mathbf{Y}_i$ and $\mathbf{Y}_{i-1}$. We shall see shortly how we may find a_i and b_i, but first we shall prove that $\mathbf{Y}_i$ is generally orthogonal to each previous vector of the sequence, and then that $\mathbf{Y}^{-1}\mathbf{A}\mathbf{Y} = \mathbf{C}$.

Theorem 8.1

If $\mathbf{Y}_{i+1}$ is orthogonal to $\mathbf{Y}_i$ and $\mathbf{Y}_{i-1}$, then it is orthogonal to each previous vector in the sequence.

Proof

From our choice of a_2 and b_2, $\mathbf{Y}_3$ will be orthogonal to $\mathbf{Y}_2$ and $\mathbf{Y}_1$. Assume that $\mathbf{Y}_i$ is orthogonal to each previous vector. Then from equation (8.1) we have

$$\mathbf{Y}_{i+1} = \mathbf{A}\mathbf{Y}_i - a_i\mathbf{Y}_i - b_i\mathbf{Y}_{i-1}$$

so that, if $r < i-1$,

$$\begin{aligned}\mathbf{Y}_r^T\mathbf{Y}_{i+1} &= \mathbf{Y}_r^T\mathbf{A}\mathbf{Y}_i - a_i\mathbf{Y}_r^T\mathbf{Y}_i - b_i\mathbf{Y}_r^T\mathbf{Y}_{i-1}\\ &= \mathbf{Y}_r^T\mathbf{A}\mathbf{Y}_i\\ &= \mathbf{Y}_i^T\mathbf{A}\mathbf{Y}_r\\ &= \mathbf{Y}_i^T(\mathbf{Y}_{r+1} + a_r\mathbf{Y}_r + b_r\mathbf{Y}_{r-1})\\ &= 0\end{aligned}$$

Hence by induction the theorem is proved.

Theorem 8.2

$\mathbf{C} = \mathbf{Y}^{-1}\mathbf{A}\mathbf{Y}$, providing that $\mathbf{Y}_i \neq 0$, $i = 1, 2, \ldots, n$.

Proof

Now $\mathbf{Y} = (\,\mathbf{Y}_1 \quad \mathbf{Y}_2 \quad \ldots \quad \mathbf{Y}_n\,)$, so that

$$\mathbf{A}\mathbf{Y} = \mathbf{A}(\,\mathbf{Y}_1 \quad \mathbf{Y}_2 \quad \ldots \quad \mathbf{Y}_n\,) = (\,\mathbf{A}\mathbf{Y}_1 \quad \mathbf{A}\mathbf{Y}_2 \quad \ldots \quad \mathbf{A}\mathbf{Y}_n\,)$$

Also

$$\mathbf{Y}\mathbf{C} = (\,\mathbf{Y}_1 \quad \mathbf{Y}_2 \quad \ldots \quad \mathbf{Y}_n\,)\begin{pmatrix} a_1 & b_2 & 0 & 0 & \ldots & 0 & 0 & 0\\ 1 & a_2 & b_3 & 0 & \ldots & 0 & 0 & 0\\ 0 & 1 & a_3 & b_4 & \ldots & 0 & 0 & 0\\ \vdots & \vdots & \vdots & \vdots & & \vdots & \vdots & \vdots\\ 0 & 0 & 0 & 0 & \ldots & 0 & 1 & a_n \end{pmatrix}$$

$$= (a_1\mathbf{Y}_1 + \mathbf{Y}_2 \; b_2\mathbf{Y}_1 + a_2\mathbf{Y}_2 + \mathbf{Y}_3 \ldots b_n\mathbf{Y}_{n-1} + a_n\mathbf{Y}_n)$$

From equation (8.1) this gives

$$\mathbf{Y}\mathbf{C} = (\,\mathbf{A}\mathbf{Y}_1 \quad \mathbf{A}\mathbf{Y}_2 \quad \ldots \quad \mathbf{A}\mathbf{Y}_n\,) = \mathbf{A}\mathbf{Y}$$

Since the sequence $\mathbf{Y}_1, \mathbf{Y}_2, \ldots, \mathbf{Y}_n$ forms an orthogonal set of vectors, providing they are all non-zero $\mathbf{Y}^{-1}$ will exist and hence

$$\mathbf{C} = \mathbf{Y}^{-1}\mathbf{A}\mathbf{Y}$$

as required.

To find a_i we premultiply equation (8.1) by $\mathbf{Y}_i^T$ which gives

$$\mathbf{Y}_i^T\mathbf{Y}_{i+1} = \mathbf{Y}_i^T\mathbf{A}\mathbf{Y}_i - a_i\mathbf{Y}_i^T\mathbf{Y}_i - b_i\mathbf{Y}_i^T\mathbf{Y}_{i-1}$$

and from the orthogonality requirement we get

$$0 = \mathbf{Y}_i^T\mathbf{A}\mathbf{Y}_i - a_i\mathbf{Y}_i^T\mathbf{Y}_i$$

or

$$a_i\mathbf{Y}_i^T\mathbf{Y}_i = \mathbf{Y}_i^T\mathbf{A}\mathbf{Y}_i \tag{8.2}$$

Since $\mathbf{Y}_i^T\mathbf{Y}_i$ and $\mathbf{Y}_i^T\mathbf{A}\mathbf{Y}_i$ are both single elements we can easily determine a_i.

To find b_i we premultiply equation (8.1) by $\mathbf{Y}_{i-1}^T$, which gives

$$\mathbf{Y}_{i-1}^T \mathbf{Y}_{i+1} = \mathbf{Y}_{i-1}^T \mathbf{A}\mathbf{Y}_i - a_i \mathbf{Y}_{i-1}^T \mathbf{Y}_i - b_i \mathbf{Y}_{i-1}^T \mathbf{Y}_{i-1}$$

so that

$$\begin{aligned} b_i \mathbf{Y}_{i-1}^T \mathbf{Y}_{i-1} &= \mathbf{Y}_{i-1}^T \mathbf{A}\mathbf{Y}_i \\ &= \mathbf{Y}_i^T \mathbf{A}\mathbf{Y}_{i-1} \\ &= \mathbf{Y}_i^T(\mathbf{Y}_i + a_{i-1}\mathbf{Y}_{i-1} + b_{i-1}\mathbf{Y}_{i-2}) \\ &= \mathbf{Y}_i^T \mathbf{Y}_i \end{aligned} \tag{8.3}$$

which enables us to find conveniently b_i. We note that if we normalize each vector $\mathbf{Y}_i$ so that $\mathbf{Y}_i^T \mathbf{Y}_i = 1$, then each $b_i = 1$ and the matrix $\mathbf{Y}$ is orthogonal.

Example 8.1

We take here the matrix $\mathbf{A}$ of example 6.1

$$\mathbf{A} = \begin{pmatrix} 0 & 3 & 4 \\ 3 & 1 & -1 \\ 4 & -1 & 0 \end{pmatrix}$$

Let

$$\mathbf{Y}_1 = \begin{pmatrix} 1 \\ 0 \\ 0 \end{pmatrix}$$

Then

$$\mathbf{Y}_1^T \mathbf{Y}_1 = (\ 1\ ); \quad \mathbf{A}\mathbf{Y}_1 = \begin{pmatrix} 0 \\ 3 \\ 4 \end{pmatrix}; \quad \mathbf{Y}_1^T \mathbf{A}\mathbf{Y}_1 = (\ 0\ )$$

so that

$$a_1 = \frac{0}{1} = 0$$

and

$$\begin{aligned} \mathbf{Y}_2 = \mathbf{A}\mathbf{Y}_1 - a_1 \mathbf{Y}_1 &= \mathbf{A}\mathbf{Y}_1 \\ &= \begin{pmatrix} 0 \\ 3 \\ 4 \end{pmatrix} \end{aligned}$$

$$\mathbf{Y}_2^T \mathbf{Y}_2 = (\ 25\ ), \quad \mathbf{A}\mathbf{Y}_2 = \begin{pmatrix} 0 & 3 & 4 \\ 3 & 1 & -1 \\ 4 & -1 & 0 \end{pmatrix} \begin{pmatrix} 0 \\ 3 \\ 4 \end{pmatrix} = \begin{pmatrix} 25 \\ -1 \\ -3 \end{pmatrix}$$

$$\mathbf{Y}_2^T \mathbf{A}\mathbf{Y}_2 = (\ -15\ )$$

so that

$$a_2 = \frac{-15}{25} = -\tfrac{3}{5} \quad \text{and} \quad b_2 = \tfrac{25}{1} = 25$$

and

$$\mathbf{Y}_3 = \mathbf{AY}_2 - a_2\mathbf{Y}_2 - b_2\mathbf{Y}_1 = \begin{pmatrix} 25 \\ -1 \\ -3 \end{pmatrix} + \frac{3}{5}\begin{pmatrix} 0 \\ 3 \\ 4 \end{pmatrix} - 25\begin{pmatrix} 1 \\ 0 \\ 0 \end{pmatrix} = \begin{pmatrix} 0 \\ \frac{4}{5} \\ -\frac{3}{5} \end{pmatrix}$$

$$\mathbf{Y}_3^T\mathbf{Y}_3 = (\ 1\ ), \quad \mathbf{AY}_3 = \begin{pmatrix} 0 & 3 & 4 \\ 3 & 1 & -1 \\ 4 & -1 & 0 \end{pmatrix}\begin{pmatrix} 0 \\ \frac{4}{5} \\ -\frac{3}{5} \end{pmatrix} = \begin{pmatrix} 0 \\ \frac{7}{5} \\ -\frac{4}{5} \end{pmatrix}$$

$$\mathbf{Y}_3^T\mathbf{AY}_3 = (\ \tfrac{8}{5}\ )$$

so that

$$a_3 = \frac{\frac{8}{5}}{1} = \frac{8}{5} \quad \text{and} \quad b_3 = \frac{1}{25}$$

Hence

$$\mathbf{C} = \begin{pmatrix} 0 & 25 & 0 \\ 1 & -\frac{3}{5} & \frac{1}{25} \\ 0 & 1 & \frac{8}{5} \end{pmatrix}$$

which is the required tridiagonal matrix. If we compare this matrix with the matrix $\mathbf{A}_1$ of example 6.1, we can see that the leading diagonals are identical, and that the product of corresponding elements on the super-diagonal of $\mathbf{A}_1$ is the same as the element on the upper superdiagonal of $\mathbf{C}$. Clearly we shall get the same Sturm sequence here as in example 6.1.

8.2 Dealing with a Zero Vector in the Lanczos Method

It is clear that theorem 8.2 does not hold if any $\mathbf{Y}_i = 0$, and, furthermore, from equations (8.1) and (8.3) we can see that all successive vectors will also be zero.

If we find that $\mathbf{Y}_i = 0$, then we can select any non-zero vector $\mathbf{X}_i$, which is orthogonal to all the previous vectors in the sequence. Then $\mathbf{Y}_{i+1}$ is given by

$$\mathbf{Y}_{i+1} = \mathbf{AX}_i - a_i\mathbf{X}_i - b_i\mathbf{Y}_{i-1} \tag{8.4}$$

Premultiplying equation (8.4) by $\mathbf{X}_i^T$ we get

$$\mathbf{X}_i^T\mathbf{Y}_{i+1} = \mathbf{X}_i^T\mathbf{AX}_i - a_i\mathbf{X}_i^T\mathbf{X}_i - b_i\mathbf{X}_i^T\mathbf{Y}_{i-1}$$

so that

$$0 = \mathbf{X}_i^T\mathbf{AX}_i - a_i\mathbf{X}_i^T\mathbf{X}_i$$

or

$$a_i \mathbf{X}_i^T \mathbf{X}_i = \mathbf{X}_i^T \mathbf{A}\mathbf{X}_i$$

which is of the same form as equation (8.2). Now premultiplying equation (8.1) by $\mathbf{Y}_{i-1}^T$ we get

$$\mathbf{Y}_{i-1}^T \mathbf{Y}_{i+1} = \mathbf{Y}_{i-1}^T \mathbf{A}\mathbf{X}_i - a_i \mathbf{Y}_{i-1}^T \mathbf{X}_i - b_i \mathbf{Y}_{i-1}^T \mathbf{Y}_{i-1}$$

so that

$$0 = \mathbf{Y}_{i-1}^T \mathbf{A}\mathbf{X}_i - b_i \mathbf{Y}_{i-1}^T \mathbf{Y}_{i-1}$$

or

$$\begin{aligned} b_i \mathbf{Y}_{i-1}^T \mathbf{Y}_{i-1} &= \mathbf{Y}_{i-1}^T \mathbf{A}\mathbf{X}_i = \mathbf{X}_i^T \mathbf{A}\mathbf{Y}_{i-1} \\ &= \mathbf{X}_i^T(\mathbf{Y}_i + a_{i-1}\mathbf{Y}_{i-1} + b_{i-1}\mathbf{Y}_{i-2}) \\ &= \mathbf{X}_i^T \mathbf{Y}_i = 0 \end{aligned}$$

since $\mathbf{Y}_i$ is zero. Since $\mathbf{Y}_{i-1}^T \mathbf{Y}_{i-1} \neq 0$, this means that $b_i = 0$, and we can see that this means that the matrix $\mathbf{C}$ can be partitioned into two tridiagonal submatrices.

Example 8.2

$$\mathbf{A} = \begin{pmatrix} 5 & -2 & -2 \\ -2 & 2 & -4 \\ -2 & -4 & 2 \end{pmatrix}, \quad \mathbf{Y}_1 = \begin{pmatrix} 1 \\ 0 \\ 0 \end{pmatrix}$$

Then

$$\mathbf{Y}_1^T \mathbf{Y}_1 = (\ 1\ ); \quad \mathbf{A}\mathbf{Y}_1 = \begin{pmatrix} 5 \\ -2 \\ -2 \end{pmatrix}; \quad \mathbf{Y}_1^T \mathbf{A}\mathbf{Y}_1 = (\ 5\ )$$

so that

$$a_1 = \tfrac{5}{1} = 5$$

and

$$\mathbf{Y}_2 = \begin{pmatrix} 5 \\ -2 \\ -2 \end{pmatrix} - 5\begin{pmatrix} 1 \\ 0 \\ 0 \end{pmatrix} = \begin{pmatrix} 0 \\ -2 \\ -2 \end{pmatrix}$$

$$\mathbf{Y}_2^T \mathbf{Y}_2 = (\ 8\ ), \quad \mathbf{A}\mathbf{Y}_2 = \begin{pmatrix} 8 \\ 4 \\ 4 \end{pmatrix}; \quad \mathbf{Y}_2^T \mathbf{A}\mathbf{Y}_2 = (\ -16\ )$$

so that

$$a_2 = -2 \quad \text{and} \quad b_2 = 8$$

which gives

$$\mathbf{Y}_3 = \begin{pmatrix} 8 \\ 4 \\ 4 \end{pmatrix} + 2\begin{pmatrix} 0 \\ -2 \\ -2 \end{pmatrix} - 8\begin{pmatrix} 1 \\ 0 \\ 0 \end{pmatrix} = \begin{pmatrix} 0 \\ 0 \\ 0 \end{pmatrix} = 0$$

Hence $b_3 = 0$. Now any vector of the form

$$\mathbf{X}^T = (\,0 \quad 0 \quad c\,)$$

will be linearly independent of $\mathbf{Y}_1$ and $\mathbf{Y}_2$ and we can then use the Gram–Schmidt process to find an orthogonal vector.† When $c = 4$, this gives

$$\mathbf{X}_3 = \begin{pmatrix} 0 \\ -2 \\ 2 \end{pmatrix}$$

so that

$$\mathbf{X}_3^T\mathbf{X}_3 = (\,8\,), \quad \mathbf{AX}_3 = \begin{pmatrix} 0 \\ -12 \\ 12 \end{pmatrix}, \quad \mathbf{X}_3^T\mathbf{AX}_3 = (\,48\,)$$

Hence

$$a_3 = 6$$

The required tridiagonal form is now given by

$$\mathbf{C} = \begin{pmatrix} 5 & 8 & 0 \\ 1 & -2 & 0 \\ 0 & 0 & 6 \end{pmatrix}$$

which has roots $\lambda_1 = \lambda_2 = 6$, $\lambda_3 = -3$.

8.3 Number of Calculations Required by Lanczos' Method

It is convenient to take as the vector $\mathbf{Y}_1$, the vector whose first component is unity and the rest zero. In this case no multiplications or divisions are required to find a_1 and $\mathbf{Y}_2$. Now to form $\mathbf{AY}_2$ will require n^2 multiplications, to form $\mathbf{Y}_2^T\mathbf{Y}_2$ n multiplications, to form $\mathbf{Y}_2^T\mathbf{AY}_2$ a further n multiplications, and to form a_2 one division. $\mathbf{Y}_3$ will require n multiplications. So in forming $\mathbf{Y}_3$ we perform

$$n^2 + 3n + 1 \text{ calculations}$$

† See reference 11, p. 442. Here, of course, we only have to perform the one step of the process since the previous vectors are already orthogonal.

To find $\mathbf{Y}_4$ requires $(n^2+3n+1)+(1+n) = n^2+4n+2$ calculations, and similarly for all succeeding vectors up to $\mathbf{Y}_n$. To find a_n and b_n requires a further n^2+2n+2 calculations. So, in all, the number of calculations needed by the Lanczos method is given by

$$\begin{aligned} S_n &= (n^2+3n+1)+(n-3)(n^2+4n+2)+(n^2+2n+2) \\ &= n^3+3n^2-5n-3 \end{aligned}$$

(See Table 8.1.)

TABLE 8.1

n	n^3+3n^2-5n-3
3	36
4	65
5	172
6	289
7	452
8	661
9	924
10	1247
20	9097
50	132 247
100	1 029 497

8.4 Further Comments on Lanczos' Method for Symmetric Matrices

Wilkinson† points out that in practice the vectors can quickly lose their orthogonality property in which case we have to reorthogonalize. If we find that $\mathbf{Y}_r$ is not strictly orthogonal to all previous vectors, we replace it by $\mathbf{X}_r$ such that

$$\mathbf{X}_r = \mathbf{Y}_r - d_1\mathbf{Y}_1 - d_2\mathbf{Y}_2 - \dots - d_{r-1}\mathbf{Y}_{r-1}$$

where

$$d_i\mathbf{Y}_i^T\mathbf{Y}_i = \mathbf{Y}_i^T\mathbf{Y}_r$$

It is, of course, essential to the Lanczos method that strict orthogonality is maintained, so that we ensure that our sequence of vectors are all linearly independent. This necessity detracts somewhat from the practical value of Lanczos' method and certainly for computer use Householder's method is preferable. We note that, having found a latent vector of $\mathbf{C}$, we only have to multiply by $\mathbf{Y}$ to find the latent vector of $\mathbf{A}$.

† See reference 7, p. 394.

8.5 The Method of Lanczos for Unsymmetric Matrices

The method of Lanczos extends conveniently to reducing an unsymmetric matrix to tridiagonal form. In this case Lanczos' method constructs two sequences of bi-orthogonal vectors.

We reduce $\mathbf{A}$ to the matrix $\mathbf{C}$ by constructing two sequences of vectors $\mathbf{Y}_1, \mathbf{Y}_2, \ldots, \mathbf{Y}_n$ and $\mathbf{Z}_1, \mathbf{Z}_2, \ldots, \mathbf{Z}_n$ such that each $\mathbf{Z}_i$ is orthogonal to the previous $\mathbf{Y}$ vectors, and each $\mathbf{Y}_i$ is orthogonal to the previous $\mathbf{Z}$ vectors.

The vectors $\mathbf{Y}_{i+1}$ and $\mathbf{Z}_{i+1}$ are given by the recurrence relations

$$\mathbf{Y}_{i+1} = \mathbf{A}\mathbf{Y}_i - a_i \mathbf{Y}_i - b_i \mathbf{Y}_{i-1}, \quad \mathbf{Z}_{i+1} = \mathbf{A}^T \mathbf{Z}_i - c_i \mathbf{Z}_i - d_i \mathbf{Z}_{i-1}$$

By the orthogonality requirements we find that

$$a_i \mathbf{Z}_i^T \mathbf{Y}_i = \mathbf{Z}_i^T \mathbf{A}\mathbf{Y}_i, \quad b_i \mathbf{Z}_{i-1}^T \mathbf{Y}_{i-1} = \mathbf{Y}_i^T \mathbf{Z}_i$$
$$c_i \mathbf{Y}_i^T \mathbf{Z}_i = \mathbf{Y}_i^T \mathbf{A}^T \mathbf{Z}_i, \quad d_i \mathbf{Y}_{i-1}^T \mathbf{Z}_{i-1} = \mathbf{Z}_i^T \mathbf{Y}_i$$

which is similar to the symmetric case. Now it is clear that

$$\mathbf{Z}_{i-1}^T \mathbf{Y}_{i-1} = \mathbf{Y}_{i-1}^T \mathbf{Z}_{i-1}, \quad \mathbf{Z}_i^T \mathbf{Y}_i = \mathbf{Y}_i^T \mathbf{Z}_i, \quad \mathbf{Z}_i^T \mathbf{A}\mathbf{Y}_i = \mathbf{Y}_i^T \mathbf{A}^T \mathbf{Z}_i$$

which means that

$$a_i = c_i \quad \text{and} \quad b_i = d_i$$

so that the equations we need are

$$\mathbf{Y}_{i+1} = \mathbf{A}\mathbf{Y}_i - a_i \mathbf{Y}_i - b_i \mathbf{Y}_{i-1} \tag{8.5}$$
$$\mathbf{Z}_{i+1} = \mathbf{A}^T \mathbf{Z}_i - a_i \mathbf{Z}_i - b_i \mathbf{Z}_{i-1} \tag{8.6}$$
$$a_i \mathbf{Z}_i^T \mathbf{Y}_i = \mathbf{Z}_i^T \mathbf{A}\mathbf{Y}_i \tag{8.7}$$
$$b_i \mathbf{Z}_{i-1}^T \mathbf{Y}_{i-1} = \mathbf{Z}_i^T \mathbf{Y}_i \tag{8.8}$$

If $\mathbf{Y}$ is the matrix whose columns are the vectors $\mathbf{Y}_1, \mathbf{Y}_2, \ldots, \mathbf{Y}_n$, and $\mathbf{Z}$ is the matrix whose columns are the vectors $\mathbf{Z}_1, \mathbf{Z}_2, \ldots, \mathbf{Z}_n$, we now proceed to show that, in general,

$$\mathbf{C} = \mathbf{Y}^{-1}\mathbf{A}\mathbf{Y} = \mathbf{Z}^{-1}\mathbf{A}^T\mathbf{Z}$$

Theorem 8.3

If $\mathbf{Y}_{i+1}$ is orthogonal to $\mathbf{Z}_i$ and $\mathbf{Z}_{i-1}$, then it is orthogonal to $\mathbf{Z}_j$ for all j such that $1 \leqslant j \leqslant i-2$. Similarly, if $\mathbf{Z}_{i+1}$ is orthogonal to $\mathbf{Y}_i$ and $\mathbf{Y}_{i-1}$, then it is orthogonal to $\mathbf{Y}_j$ for all j such that $1 \leqslant j \leqslant i-2$.

Proof

The proof here follows exactly the same lines as that of theorem 8.1 and it would be mere repetition to include it.

Theorem 8.4

$\mathbf{Y}^{-1}$ and $\mathbf{Z}^{-1}$ both exist providing that $\mathbf{Z}_i^T \mathbf{Y}_i \neq 0$ for all i.

Proof

Now

$$\mathbf{Z}^T\mathbf{Y} = \begin{pmatrix} \mathbf{Z}_1^T \\ \mathbf{Z}_2^T \\ \vdots \\ \mathbf{Z}_n^T \end{pmatrix} (\mathbf{Y}_1 \quad \mathbf{Y}_2 \quad \dots \quad \mathbf{Y}_n) = \begin{pmatrix} \mathbf{Z}_1^T\mathbf{Y}_1 & \mathbf{Z}_1^T\mathbf{Y}_2 & \dots & \mathbf{Z}_1^T\mathbf{Y}_n \\ \mathbf{Z}_2^T\mathbf{Y}_1 & \mathbf{Z}_2^T\mathbf{Y}_2 & \dots & \mathbf{Z}_2^T\mathbf{Y}_n \\ \vdots & \vdots & & \vdots \\ \mathbf{Z}_n^T\mathbf{Y}_1 & \mathbf{Z}_n^T\mathbf{Y}_2 & \dots & \mathbf{Z}_n^T\mathbf{Y}_n \end{pmatrix}$$

$$= \begin{pmatrix} \mathbf{Z}_1^T\mathbf{Y}_1 & 0 & \dots & 0 \\ 0 & \mathbf{Z}_2^T\mathbf{Y}_2 & \dots & 0 \\ \vdots & \vdots & & \vdots \\ 0 & 0 & \dots & \mathbf{Z}_n^T\mathbf{Y}_n \end{pmatrix}$$

so that providing $\mathbf{Z}_i^T\mathbf{Y}_i \neq 0$ for all i it is clear that $(\mathbf{Z}^T\mathbf{Y})^{-1}$ exists and hence $|\mathbf{Z}^T\mathbf{Y}| \neq 0$. This, of course, means that $|\mathbf{Z}^T| = |\mathbf{Z}| \neq 0$ and $|\mathbf{Y}| \neq 0$. Hence $\mathbf{Z}^{-1}$ and $\mathbf{Y}^{-1}$ both exist, as required.

Theorem 8.5

$\mathbf{C} = \mathbf{Y}^{-1}\mathbf{A}\mathbf{Y} = \mathbf{Z}^{-1}\mathbf{A}^T\mathbf{Z}$ providing that $\mathbf{Z}_i^T\mathbf{Y}_i \neq 0$ for all i.

Proof

Theorem 8.4 has established the existence of $\mathbf{Y}^{-1}$ and $\mathbf{Z}^{-1}$. This proof is now identical with that of theorem 8.2.

We note that if we choose $\mathbf{Y}_i$ and $\mathbf{Z}_i$ so that $\mathbf{Z}_i^T\mathbf{Y}_i = 1$, then $b_i = 1$, and from theorem 8.4 we see that $\mathbf{Z}^T\mathbf{Y} = \mathbf{I}$. Hence $\mathbf{Z}^T = \mathbf{Y}^{-1}$.

Example 8.3

Here we take the matrix $\mathbf{A}$ of example 3.1.

$$\mathbf{A} = \begin{pmatrix} 0 & -2 & 5 \\ -7 & 1 & 9 \\ -1 & -2 & 6 \end{pmatrix}$$

Let

$$\mathbf{Y}_1 = \mathbf{Z}_1 = \begin{pmatrix} 1 \\ 0 \\ 0 \end{pmatrix}$$

$$\mathbf{A}\mathbf{Y}_1 = \begin{pmatrix} 0 \\ -7 \\ -1 \end{pmatrix}, \quad \mathbf{A}^T\mathbf{Z}_1 = \begin{pmatrix} 0 \\ -2 \\ 5 \end{pmatrix}$$

$$\mathbf{Z}_1^T\mathbf{Y}_1 = (1), \quad \mathbf{Z}_1^T\mathbf{A}\mathbf{Y}_1 = (0)$$

so that

$$a_1 = \frac{0}{1} = 0$$

and

$$\mathbf{Y}_2 = \mathbf{AY}_1 - a_1\mathbf{Y}_1 = \begin{pmatrix} 0 \\ -7 \\ -1 \end{pmatrix}, \quad \mathbf{Z}_2 = \mathbf{A}^T\mathbf{Z}_1 - a_1\mathbf{Z}_1 = \begin{pmatrix} 0 \\ -2 \\ 5 \end{pmatrix}$$

$$\mathbf{AY}_2 = \begin{pmatrix} 9 \\ -16 \\ 8 \end{pmatrix}, \quad \mathbf{A}^T\mathbf{Z}_2 = \begin{pmatrix} 9 \\ -12 \\ 12 \end{pmatrix}$$

$$\mathbf{Z}_2^T\mathbf{Y}_2 = (\ 9\ ), \quad \mathbf{Z}_2^T\mathbf{AY}_2 = (\ 72\ )$$

so that

$$a_2 = \tfrac{72}{9} = 8, \quad b_2 = \tfrac{9}{1} = 9$$

and

$$\mathbf{Y}_3 = \begin{pmatrix} 9 \\ -16 \\ 8 \end{pmatrix} - 8\begin{pmatrix} 0 \\ -7 \\ -1 \end{pmatrix} - 9\begin{pmatrix} 1 \\ 0 \\ 0 \end{pmatrix} = \begin{pmatrix} 0 \\ 40 \\ 16 \end{pmatrix}$$

$$\mathbf{Z}_3 = \begin{pmatrix} 9 \\ -12 \\ 12 \end{pmatrix} - 8\begin{pmatrix} 0 \\ -2 \\ 5 \end{pmatrix} - 9\begin{pmatrix} 1 \\ 0 \\ 0 \end{pmatrix} = \begin{pmatrix} 0 \\ 4 \\ -28 \end{pmatrix}$$

$$\mathbf{AY}_3 = \begin{pmatrix} 0 \\ 184 \\ 16 \end{pmatrix}$$

($\mathbf{A}^T\mathbf{Z}_3$ is not required.)

$$\mathbf{Z}_3^T\mathbf{Y}_3 = (\ -288\ ), \quad \mathbf{Z}_3^T\mathbf{AY}_3 = (\ 288\ )$$

so that

$$a_3 = \frac{288}{-288} = -1 \quad \text{and} \quad b_3 = \frac{-288}{9} = -32$$

Hence the required tridiagonal matrix is

$$\mathbf{C} = \begin{pmatrix} 0 & 9 & 0 \\ 1 & 8 & -32 \\ 0 & 1 & -1 \end{pmatrix}$$

This is the matrix whose latent roots were found using Muller's method in example 5.4.

8.6 Dealing with Zero Vectors in the Unsymmetric Case

If either $\mathbf{Y}_i = 0$ or $\mathbf{Z}_i = 0$, then clearly $\mathbf{Z}_i^T \mathbf{Y}_i = 0$ and theorem 8.5 does not hold. In this case we proceed as for a zero vector in the symmetric case, that is, we choose any non-zero orthogonal vector. But in this case we have to note that, in general, $b_i \neq d_i$ unless both $\mathbf{Y}_i$ and $\mathbf{Z}_i$ are zero, which we shall now show. Suppose that $\mathbf{Y}_i = 0$ and we replace it by a non-zero vector $\mathbf{X}_i$ which is orthogonal to all the previous $\mathbf{Z}$ vectors. Then

$$\mathbf{Y}_{i+1} = \mathbf{A}\mathbf{X}_i - a_i \mathbf{X}_i - b_i \mathbf{Y}_{i-1}$$

and premultiplying by $\mathbf{Z}_{i-1}^T$ we get

$$\mathbf{Z}_{i-1}^T \mathbf{Y}_{i+1} = \mathbf{Z}_{i-1}^T \mathbf{A}\mathbf{X}_i - a_i \mathbf{Z}_{i-1}^T \mathbf{X}_i - b_i \mathbf{Z}_{i-1}^T \mathbf{Y}_{i-1}$$

so that

$$\begin{aligned} b_i \mathbf{Z}_{i-1}^T \mathbf{Y}_{i-1} &= \mathbf{Z}_{i-1}^T \mathbf{A}\mathbf{X}_i = \mathbf{X}_i^T \mathbf{A}^T \mathbf{Z}_{i-1} \\ &= \mathbf{X}_i^T (\mathbf{Z}_i + a_{i-1} \mathbf{Z}_{i-1} + b_{i-1} \mathbf{Z}_{i-2}) \\ &= \mathbf{X}_i^T \mathbf{Z}_i \end{aligned}$$

On the other hand,

$$\mathbf{Z}_{i+1} = \mathbf{A}^T \mathbf{Z}_i - c_i \mathbf{Z}_i - d_i \mathbf{Z}_{i-1}$$

and premultiplying by $\mathbf{Y}_{i-1}^T$ we get

$$\mathbf{Y}_{i-1}^T \mathbf{Z}_{i+1} = \mathbf{Y}_{i-1}^T \mathbf{A}^T \mathbf{Z}_i - c_i \mathbf{Y}_{i-1}^T \mathbf{Z}_i - d_i \mathbf{Y}_{i-1}^T \mathbf{Z}_{i-1}$$

so that

$$\begin{aligned} d_i \mathbf{Y}_{i-1}^T \mathbf{Z}_{i-1} &= \mathbf{Y}_{i-1}^T \mathbf{A}^T \mathbf{Z}_i = \mathbf{Z}_i^T \mathbf{A}\mathbf{Y}_{i-1} \\ &= \mathbf{Z}_i^T (\mathbf{Y}_i + a_{i-1} \mathbf{Y}_{i-1} + b_{i-1} \mathbf{Y}_{i-2}) \\ &= \mathbf{Z}_i^T \mathbf{Y}_i = 0 \end{aligned}$$

Hence, in general, $d_i = 0$ but $b_i \neq 0$. We further note that $\mathbf{Y}^{-1}\mathbf{A}\mathbf{Y}$ is now of the form

$$\mathbf{C} = \begin{pmatrix} a_1 & b_2 & 0 & \dots & 0 & 0 & 0 & 0 & \dots & 0 & 0 & 0 \\ 1 & a_2 & b_3 & \dots & 0 & 0 & 0 & 0 & \dots & 0 & 0 & 0 \\ \vdots & \vdots & \vdots & & \vdots & \vdots & \vdots & \vdots & & \vdots & \vdots & \vdots \\ 0 & 0 & 0 & \dots & 1 & a_{i-1} & b_i & 0 & \dots & 0 & 0 & 0 \\ 0 & 0 & 0 & \dots & 0 & 0 & a_i & b_{i+1} & \dots & 0 & 0 & 0 \\ \vdots & \vdots & \vdots & & \vdots & \vdots & \vdots & \vdots & & \vdots & \vdots & \vdots \\ 0 & 0 & 0 & \dots & 0 & 0 & 0 & 0 & \dots & 0 & 1 & a_n \end{pmatrix}$$

whereas $\mathbf{Z}^{-1}\mathbf{A}^T\mathbf{Z}$ is of the form

$$\mathbf{C}_1 = \begin{pmatrix} a_1 & b_2 & 0 & \dots & 0 & 0 & 0 & 0 & \dots & 0 & 0 & 0 \\ 1 & a_2 & b_3 & \dots & 0 & 0 & 0 & 0 & \dots & 0 & 0 & 0 \\ \vdots & \vdots & \vdots & & \vdots & \vdots & \vdots & \vdots & & \vdots & \vdots & \vdots \\ 0 & 0 & 0 & \dots & 1 & a_{i-1} & 0 & 0 & \dots & 0 & 0 & 0 \\ 0 & 0 & 0 & \dots & 0 & 1 & a_i & b_{i+1} & \dots & 0 & 0 & 0 \\ \vdots & \vdots & \vdots & & \vdots & \vdots & \vdots & \vdots & & \vdots & \vdots & \vdots \\ 0 & 0 & 0 & \dots & 0 & 0 & 0 & 0 & \dots & 0 & 1 & a_n \end{pmatrix}$$

Although $\mathbf{C} \neq \mathbf{C}_1$ we can partition them both into the same tridiagonal submatrices.

Example 8.4

Here we take the matrix $\mathbf{A}$ of example 3.3.

$$\mathbf{A} = \begin{pmatrix} 1 & -1 & 3 & 4 \\ 4 & 1 & 2 & 1 \\ 4 & 2 & 1 & -1 \\ 0 & -1 & 1 & 0 \end{pmatrix}, \quad \mathbf{Y}_1 = \mathbf{Z}_1 = \begin{pmatrix} 1 \\ 0 \\ 0 \\ 0 \end{pmatrix}$$

$$\mathbf{AY}_1 = \begin{pmatrix} 1 \\ 4 \\ 4 \\ 0 \end{pmatrix}, \quad \mathbf{A}^T\mathbf{Z}_1 = \begin{pmatrix} 1 \\ -1 \\ 3 \\ 4 \end{pmatrix}$$

$$\mathbf{Z}_1^T\mathbf{Y}_1 = (\ 1\ ), \quad \mathbf{Z}_1^T\mathbf{AY}_1 = (\ 1\ )$$

$$a_1 = \tfrac{1}{1} = 1$$

$$\mathbf{Y}_2 = \begin{pmatrix} 0 \\ 4 \\ 4 \\ 0 \end{pmatrix}, \quad \mathbf{Z}_2 = \begin{pmatrix} 0 \\ -1 \\ 3 \\ 4 \end{pmatrix}$$

$$\mathbf{AY}_2 = \begin{pmatrix} 8 \\ 12 \\ 12 \\ 0 \end{pmatrix}, \quad \mathbf{A}^T\mathbf{Z}_2 = \begin{pmatrix} 8 \\ 1 \\ 5 \\ -4 \end{pmatrix}$$

$$\mathbf{Z}_2^T\mathbf{Y}_2 = (\ 8\ ), \quad \mathbf{Z}_2^T\mathbf{AY}_2 = (\ 24\ )$$

$$a_2 = \tfrac{24}{8} = 3, \quad b_2 = \tfrac{8}{1} = 8$$

$$\mathbf{Y}_3 = \begin{pmatrix} 0 \\ 0 \\ 0 \\ 0 \end{pmatrix}, \quad \mathbf{Z}_3 = \begin{pmatrix} 0 \\ 4 \\ -4 \\ -16 \end{pmatrix}$$

Since $\mathbf{Y}_3 = 0$ we have to choose a non-zero vector $\mathbf{X}_3$ that is orthogonal to both $\mathbf{Z}_1$ and $\mathbf{Z}_2$. Any vector of the form

$$\mathbf{X}^T = (\,0 \quad 0 \quad 0 \quad c\,)$$

where $c \neq 0$ is linearly independent of $\mathbf{Z}_1$ and $\mathbf{Z}_2$. When $c = 13$ the Gram–Schmidt process gives

$$\mathbf{X}_3^T = (\,0 \quad 2 \quad -6 \quad 5\,)$$

Since $\mathbf{Y}_3 = 0$, $d_3 = 0$.

$$\mathbf{AX}_3 = \begin{pmatrix} 0 \\ -5 \\ -7 \\ -8 \end{pmatrix}, \quad \mathbf{A}^T\mathbf{Z}_3 = \begin{pmatrix} 0 \\ 12 \\ -12 \\ 8 \end{pmatrix}$$

$$\mathbf{Z}_3^T\mathbf{X}_3 = (\,-48\,), \quad \mathbf{Z}_3^T\mathbf{AX}_3 = (\,136\,)$$

$$a_3 = \frac{136}{-48} = -\frac{17}{6}, \quad b_3 = -\frac{48}{8} = -6$$

$$\mathbf{Y}_4 = \mathbf{AX}_3 - a_3\mathbf{X}_3 - b_3\mathbf{Y}_2 = \begin{pmatrix} 0 \\ \frac{74}{3} \\ 0 \\ \frac{37}{6} \end{pmatrix}$$

$$\mathbf{Z}_4 = \mathbf{A}^T\mathbf{Z}_3 - a_3\mathbf{Z}_3 - d_3\mathbf{Z}_2 = \begin{pmatrix} 0 \\ \frac{70}{3} \\ -\frac{70}{3} \\ -\frac{112}{3} \end{pmatrix}$$

$$\mathbf{AY}_4 = \begin{pmatrix} 0 \\ \frac{185}{6} \\ \frac{259}{6} \\ -\frac{74}{3} \end{pmatrix}, \quad \mathbf{Z}_4^T\mathbf{Y}_4 = (\,\tfrac{1036}{3}\,), \quad \mathbf{Z}_4^T\mathbf{AY}_4 = (\,\tfrac{5698}{9}\,)$$

$$a_4 = \tfrac{11}{6}, \quad b_4 = -\tfrac{259}{36}$$

We can use either of the two tridiagonal matrices

$$\mathbf{C} = \begin{pmatrix} 1 & 8 & 0 & 0 \\ 1 & 3 & -6 & 0 \\ 0 & 0 & -\frac{17}{6} & -\frac{259}{36} \\ 0 & 0 & 1 & \frac{11}{6} \end{pmatrix} \quad \text{or} \quad \mathbf{C}_1 = \begin{pmatrix} 1 & 8 & 0 & 0 \\ 1 & 3 & 0 & 0 \\ 0 & 1 & -\frac{17}{6} & -\frac{259}{36} \\ 0 & 0 & 1 & \frac{11}{6} \end{pmatrix}$$

to find the latent roots.†

8.7 Failure of Lanczos' Method for Unsymmetric Matrices

We have seen in the previous section how we can deal with the case when $\mathbf{Z}_i^T\mathbf{Y}_i = 0$ because either $\mathbf{Y}_i = 0$ or $\mathbf{Z}_i = 0$. Unfortunately it is quite possible for $\mathbf{Z}_i^T\mathbf{Y}_i = 0$, but $\mathbf{Y}_i \neq 0$ and $\mathbf{Z}_i \neq 0$. If this is so Lanczos' method breaks down because we are unable to determine a_i or b_{i+1}, as can be seen from equations (8.7) and (8.8). We note that this case is not possible for symmetric matrices with real coefficients.

Example 8.5

$$\mathbf{A} = \begin{pmatrix} 1 & -1 & 1 \\ -1 & 0 & 1 \\ -1 & 2 & 1 \end{pmatrix}, \quad \mathbf{Y}_1 = \mathbf{Z}_1 = \begin{pmatrix} 1 \\ 0 \\ 0 \end{pmatrix}$$

$$\mathbf{AY}_1 = \begin{pmatrix} 1 \\ -1 \\ -1 \end{pmatrix}, \quad \mathbf{A}^T\mathbf{Z}_1 = \begin{pmatrix} 1 \\ -1 \\ 1 \end{pmatrix}$$

$$\mathbf{Z}_1^T\mathbf{Y}_1 = (\,1\,), \quad \mathbf{Z}_1^T\mathbf{AY}_1 = (\,1\,)$$

so that

$$a_1 = \tfrac{1}{1} = 1$$

and

$$\mathbf{Y}_2 = \begin{pmatrix} 0 \\ -1 \\ -1 \end{pmatrix}, \quad \mathbf{Z}_2 = \begin{pmatrix} 0 \\ -1 \\ 1 \end{pmatrix}$$

$$\mathbf{Z}_2^T\mathbf{Y}_2 = (\,0 \quad -1 \quad 1\,)\begin{pmatrix} 0 \\ -1 \\ -1 \end{pmatrix} = (\,0\,)$$

So here $\mathbf{Y}_2 \neq 0$, $\mathbf{Z}_2 \neq 0$ but $\mathbf{Z}_2^T\mathbf{Y}_2 = 0$.

† $\mathbf{C}$ is similar to $\mathbf{A}$ and $\mathbf{C}_1$ to $\mathbf{A}^T$, so by theorem 1.9 their latent roots are the same. To find the latent vectors of $\mathbf{A}$ we must of course use $\mathbf{C}$. We note that Lanczos' method also allows us conveniently to find the latent vectors of $\mathbf{A}^T$.

In this case the only possibility is to take different initial vectors and hope that this condition is then avoided.

Here, if we take,

$$\mathbf{Y}_1 = \mathbf{Z}_1 = \begin{pmatrix} 1 \\ 1 \\ 1 \end{pmatrix}$$

we get

$$\mathbf{AY}_1 = \begin{pmatrix} 1 \\ 0 \\ 2 \end{pmatrix}, \quad \mathbf{A}^T\mathbf{Z}_1 = \begin{pmatrix} -1 \\ 1 \\ 3 \end{pmatrix}$$

$$\mathbf{Z}_1^T\mathbf{Y}_1 = (\ 3\ ), \quad \mathbf{Z}_1^T\mathbf{AY}_1 = (\ 3\ )$$

$$a_1 = \tfrac{3}{3} = 1$$

$$\mathbf{Y}_2 = \begin{pmatrix} 0 \\ -1 \\ 1 \end{pmatrix}, \quad \mathbf{Z}_2 = \begin{pmatrix} -2 \\ 0 \\ 2 \end{pmatrix}$$

$$\mathbf{AY}_2 = \begin{pmatrix} 2 \\ 1 \\ -1 \end{pmatrix}, \quad \mathbf{A}^T\mathbf{Z}_2 = \begin{pmatrix} -4 \\ 6 \\ 0 \end{pmatrix}$$

$$\mathbf{Z}_2^T\mathbf{Y}_2 = (\ 2\ ), \quad \mathbf{Z}_2^T\mathbf{AY}_2 = (\ -6\ )$$

$$a_2 = -\tfrac{6}{2} = -3, \quad b_2 = \tfrac{2}{3}$$

$$\mathbf{Y}_3 = \begin{pmatrix} \tfrac{4}{3} \\ -\tfrac{8}{3} \\ \tfrac{4}{3} \end{pmatrix}, \quad \mathbf{Z}_3 = \begin{pmatrix} -\tfrac{32}{3} \\ \tfrac{16}{3} \\ \tfrac{16}{3} \end{pmatrix}$$

$$\mathbf{AY}_3 = \begin{pmatrix} \tfrac{16}{3} \\ 0 \\ -\tfrac{16}{3} \end{pmatrix}, \quad \mathbf{Z}_3^T\mathbf{Y}_3 = (\ -\tfrac{64}{3}\ ), \quad \mathbf{Z}_3^T\mathbf{AY}_3 = (\ -\tfrac{256}{3}\ )$$

$$a_3 = 4, \quad b_3 = -\tfrac{32}{3}$$

So finally the required tridiagonal matrix is

$$\mathbf{C} = \begin{pmatrix} 1 & \frac{2}{3} & 0 \\ 1 & -3 & -\frac{32}{3} \\ 0 & 1 & 4 \end{pmatrix}$$

8.8 Relationship Between the Lanczos and Krylov Methods

It is instructive to investigate the circumstances under which either $\mathbf{Y}_i$ or $\mathbf{Z}_i$ will be zero, which we now do. From equation (8.5) we have

$$\mathbf{Y}_{i+1} = \mathbf{A}\mathbf{Y}_i - a_i \mathbf{Y}_i - b_i \mathbf{Y}_{i-1}$$

so that

$$\mathbf{Y}_2 = \mathbf{A}\mathbf{Y}_1 - a_1 \mathbf{Y}_1 = (\mathbf{A} - a_1 \mathbf{I})\,\mathbf{Y}_1$$

$$\begin{aligned} \mathbf{Y}_3 &= \mathbf{A}\mathbf{Y}_2 - a_2 \mathbf{Y}_2 - b_2 \mathbf{Y}_1 \\ &= (\mathbf{A} - a_2 \mathbf{I})(\mathbf{A} - a_1 \mathbf{I})\,\mathbf{Y}_1 - b_2 \mathbf{Y}_1 \\ &= [\mathbf{A}^2 - (a_1 + a_2)\,\mathbf{A} + (a_1 a_2 - b_2)\,\mathbf{I}]\,\mathbf{Y}_1 \end{aligned}$$

Clearly, in general, we may put

$$\begin{aligned} \mathbf{Y}_{r+1} &= [\mathbf{A}^r + p_1 \mathbf{A}^{r-1} + p_2 \mathbf{A}^{r-2} + \ldots + p_{r-1} \mathbf{A} + p_r \mathbf{I}]\,Y_1 \\ &= g(\mathbf{A})\,\mathbf{Y}_1 \end{aligned}$$

where

$$g(\lambda) = \lambda^r + p_1 \lambda^{r-1} + \ldots + p_{r-1}\lambda + p_r$$

and if $\mathbf{Y}_{r+1} = 0$ then $g(\lambda)$ is the minimal polynomial of $\mathbf{Y}_1$ with respect to $\mathbf{A}$. We saw in § 4.1 that this was the polynomial found by Krylov's method. If we look back to example 8.4 we can see that the grade of $\mathbf{Y}_1$ is two, the result we had already obtained in example 4.2 using Krylov's method.

Clearly, if the grade of $\mathbf{Y}_1$ with respect to $\mathbf{A}$ is r, then we shall have $\mathbf{Y}_{r+1} = 0$. Similarly, if the grade of $\mathbf{Z}_1$ with respect to $\mathbf{A}^T$ is q, then $\mathbf{Z}_{q+1} = 0$. It is interesting to note that having $g(\mathbf{A})\,\mathbf{Y}_1 = 0$ does not necessarily imply that $g(\mathbf{A}^T)\,\mathbf{Y}_1 = 0$.†

We now have

$$\mathbf{Y}_{r+1} = g(\mathbf{A})\,\mathbf{Y}_1$$

and

$$\mathbf{Z}_{r+1} = g(\mathbf{A}^T)\,\mathbf{Z}_1$$

† See example 8.4, where $\mathbf{Y}_3 = 0$ but $\mathbf{Z}_3 \neq 0$ although $\mathbf{Y}_1 = \mathbf{Z}_1$.

so that

$$\mathbf{Z}_{r+1}^T \mathbf{Y}_{r+1} = [g(\mathbf{A}^T)\mathbf{Z}_1]^T g(\mathbf{A})\mathbf{Y}_1$$

$$= \mathbf{Z}_1^T g(\mathbf{A}) g(\mathbf{A}) \mathbf{Y}_1$$

We can see that there are several possibilities for the vanishing of $\mathbf{Z}_{r+1}^T \mathbf{Y}_{r+1}$ without $\mathbf{Z}_{r+1}$ or $\mathbf{Y}_{r+1}$ being themselves zero.

8.9 Number of Calculations Required by Lanczos' Method

If we take as the vectors $\mathbf{Y}_1$ and $\mathbf{Z}_1$ the vector whose first component is unity and the rest zero, then no multiplications are required to find $\mathbf{Y}_2$, $\mathbf{Z}_2$, and a_1. To form $\mathbf{AY}_2$ and $\mathbf{A}^T\mathbf{Z}_2$ requires $2n^2$ multiplications. To form $\mathbf{Z}_2^T\mathbf{Y}_2$ and $\mathbf{Z}_2^T\mathbf{AY}_2$ needs a further $2n$ multiplications, and a_2 requires one division. To find $\mathbf{Y}_3$ and $\mathbf{Z}_3$ requires $2n$ multiplications. So in forming $\mathbf{Y}_3$ and $\mathbf{Z}_3$ we perform

$$2n^2+2n+2+2n = 2(n^2+2n+1) \text{ calculations}$$

To find $\mathbf{Y}_4$ and $\mathbf{Z}_4$ requires $(2n^2+4n+2)+(2n+1) = 2n^2+6n+3$ calculations, and similarly for all succeeding vectors up to $\mathbf{Y}_n$ and $\mathbf{Z}_n$. To find a_n and b_n requires a further n^2+2n+2 calculations. So in all, the number of calculations needed by the Lanczos method for unsymmetric matrices is given by

$$\mathbf{S}_n = 2(n^2+2n+1)+(n-3)(2n^2+6n+3)+n^2+2n+2$$

$$= 2n^3+3n^2-9n+4$$

$$= (n-1)(2n^2+5n-4)$$

(See Table 8.2.)

Table 8.2

n	$(n-1)(2n^2+5n-4)$
3	58
4	144
5	284
6	490
7	774
8	1148
9	1624
10	2214
20	17 024
50	257 054
100	2 029 104

8.10 Further Comments on the Lanczos Method for Unsymmetric Matrices

As in the symmetric case, in practice we shall have to reorthogonalize our vectors. Bearing this in mind, plus the possibility of zero vectors and of a complete breakdown, we can see that the method of Lanczos hardly compares with that of Danilevsky. Even in the straightforward case Lanczos' requires more than double the calculations of Danilevsky's. Nevertheless the Lanczos method is extremely interesting from a theoretical point of view and is useful to the understanding of the many related methods. Householder presents an interesting account of Lanczos' method in this direction.†

There are many other methods of reducing a matrix to tridiagonal form, most of them practically superior to Lanczos.‡

8.11 Exercises

8.1. Use the method of Lanczos to reduce the following symmetric matrices to tridiagonal form,

$$\text{(i)}\quad \mathbf{A} = \begin{pmatrix} 6 & 3 & 4 \\ 3 & 6 & 4 \\ 4 & 4 & 5 \end{pmatrix}, \qquad \text{(ii)}\quad \mathbf{A} = \begin{pmatrix} 2 & 4 & 3 & 0 \\ 4 & -2 & -3 & 2 \\ 3 & -3 & 6 & 4 \\ 0 & 2 & 4 & 3 \end{pmatrix}$$

$$\text{(iii)}\quad \mathbf{A} = \begin{pmatrix} -1 & 1 & 2 & 2 \\ 1 & 3 & -4 & -4 \\ 2 & -4 & -3 & -3 \\ 2 & -4 & -3 & -3 \end{pmatrix}$$

8.2. Complete the proofs of theorems 8.3 and 8.5.

8.3. Use the method of Lanczos to find the latent roots and vectors of the following matrices. Also find the latent vectors of the transposed matrices.

$$\text{(i)}\quad \mathbf{A} = \begin{pmatrix} 4 & 18 & 12 \\ 1 & -49 & -44 \\ -1 & 57 & 51 \end{pmatrix}, \qquad \text{(ii)}\quad \mathbf{A} = \begin{pmatrix} 4 & -0{\cdot}4 & -0{\cdot}8 \\ 1 & 0{\cdot}6 & 0{\cdot}7 \\ 2 & -0{\cdot}8 & 2{\cdot}4 \end{pmatrix}$$

$$\text{(iii)}\quad \mathbf{A} = \begin{pmatrix} 1 & 3 & -3 \\ 5 & -13 & 15 \\ 4 & -12 & 14 \end{pmatrix}, \qquad \text{(iv)}\quad \mathbf{A} = \begin{pmatrix} 1 & 2 & -1 \\ 3 & 2 & 1 \\ 6 & 11 & 3 \end{pmatrix}$$

† See reference 12.

‡ See reference 7.

8.4. Use the method of Lanczos, taking

$$\mathbf{Y}_1 = \mathbf{Z}_1 = \begin{pmatrix} 1 \\ 0 \\ 0 \end{pmatrix}$$

to show that $\mathbf{A} = \mathbf{YCY}^{-1}$ where

$$\mathbf{A} = \begin{pmatrix} 99 & 700 & -70 \\ -14 & -99 & 10 \\ 1 & 7 & 6 \end{pmatrix}, \qquad \mathbf{Y} = \begin{pmatrix} 1 & 0 & 0 \\ 0 & -14 & \frac{36}{47} \\ 0 & 1 & \frac{360}{47} \end{pmatrix}$$

$$\mathbf{C} = \begin{pmatrix} 99 & -9870 & 0 \\ 1 & -\frac{4684}{47} & \frac{684}{2209} \\ 0 & 1 & \frac{313}{47} \end{pmatrix}$$

Note that the characteristic equation of $\mathbf{C}$ is

$$(\lambda-6)(\lambda-1)(\lambda+1) = 0$$

(See exercise 3.2.)

Show that if each arithmetic calculation is performed correct only to four significant figures then the method of Lanczos yields

$$\mathbf{Y} = \begin{pmatrix} 1 & 0 & 0 \\ 0 & -14 & 1 \\ 0 & 1 & 7{\cdot}66 \end{pmatrix}, \qquad \mathbf{C}_1 = \begin{pmatrix} 99 & -9870 & 0 \\ 1 & -99{\cdot}66 & 0{\cdot}3103 \\ 0 & 1 & 0{\cdot}6627 \end{pmatrix}$$

and that the characteristic equation of $\mathbf{C}_1$ is

$$\lambda^3 - 0{\cdot}0027\lambda^2 + 2{\cdot}9123\lambda + 28{\cdot}2942 = 0$$

which has the approximate roots

$$\lambda_1 = -2{\cdot}7, \quad \lambda_2, \lambda_3 = 1{\cdot}4 \pm 2{\cdot}9i$$

Note that in this case $\mathbf{Y}_3$ is far from being orthogonal to Z_2.

Chapter 9

AN ITERATIVE METHOD FOR FINDING THE LATENT ROOT OF LARGEST MODULUS AND CORRESPONDING LATENT VECTOR

When we require all the latent roots and vectors of a matrix we would generally choose a direct method of solution. Frequently we do not require the complete solution, and in this case we are likely to choose an iterative method of solution. For example, in §§ 2.2 and 2.3 we were concerned with finding whether or not $\mathbf{A}^n \to 0$ as $n \to \infty$, and hence were interested only in the latent root of largest modulus. The method now to be described will generally find this for us. It is a slight variation from the method usually given.

9.1 The Iterative Method

Let λ_i, $i = 1, 2, \ldots, n$, be the latent roots of $\mathbf{A}$ with

$$|\lambda_1| = |\lambda_2| = \ldots = |\lambda_r|, \quad |\lambda_1| > |\lambda_{r+1}| \geqslant \ldots \geqslant |\lambda_n|$$

$\lambda_1, \lambda_2, \ldots, \lambda_r$ being real. Also, let $\mathbf{Y}_0$ be an arbitrary column vector that can be expressed in the form

$$\mathbf{Y}_0 = a_1 \mathbf{X}_1 + a_2 \mathbf{X}_2 + \ldots + a_n \mathbf{X}_n \tag{9.1}$$

where $\mathbf{X}_i$ is the latent vector associated with λ_i and $a_1 \neq 0$. We then form the sequence of vectors given by

$$\mathbf{Y}_i = k_{i-1} \mathbf{A}^2 \mathbf{Y}_{i-1}$$

k_{i-1} being for the moment an arbitrary scalar, so that

$$\mathbf{Y}_1 = k_0 \mathbf{A}^2 \mathbf{Y}_0 = k_0(a_1 \mathbf{A}^2 \mathbf{X}_1 + \ldots + a_r \mathbf{A}^2 \mathbf{X}_r + a_{r+1} \mathbf{A}^2 \mathbf{X}_{r+1} + \ldots + a_n \mathbf{A}^2 \mathbf{X}_n)$$

$$= k_0(a_1 \lambda_1^2 \mathbf{X}_1 + \ldots + a_r \lambda_r^2 \mathbf{X}_r + a_{r+1} \lambda_{r+1}^2 \mathbf{X}_{r+1} + \ldots + a_n \lambda_n^2 \mathbf{X}_n)$$

$$\mathbf{Y}_2 = k_1 \mathbf{A}^2 \mathbf{Y}_1 = k_0 k_1 \mathbf{A}^4 \mathbf{Y}_0$$

$$= k_0 k_1(a_1 \lambda_1^4 \mathbf{X}_1 + \ldots + a_r \lambda_r^4 \mathbf{X}_r + a_{r+1} \lambda_{r+1}^4 \mathbf{X}_{r+1} + \ldots + a_n \lambda_n^4 \mathbf{X}_n)$$

and in general

$$\mathbf{Y}_p = k_{p-1}\mathbf{A}^2\mathbf{Y}_{p-1} = k_0 k_1 \ldots k_{p-1}\mathbf{A}^{2p}\mathbf{Y}_0$$

$$= k_0 k_1 \ldots k_{p-1}(a_1\lambda_1^{2p}\mathbf{X}_1 + \ldots + a_r\lambda_r^{2p}\mathbf{X}_r + a_{r+1}\lambda_{r+1}^{2p}\mathbf{X}_{r+1} + \ldots + a_n\lambda_n^{2p}\mathbf{X}_n)$$

Now $\lambda_1^2 = \lambda_2^2 = \ldots = \lambda_r^2$, so that

$$\mathbf{Y}_p = k_0 k_1 \ldots k_{p-1}\lambda_1^{2p}\left\{(a_1\mathbf{X}_1 + \ldots + a_r\mathbf{X}_r) + \left[a_{r+1}\left(\frac{\lambda_{r+1}}{\lambda_1}\right)^{2p}\mathbf{X}_{r+1} + \ldots + a_n\left(\frac{\lambda_n}{\lambda_1}\right)^{2p}\mathbf{X}_n\right]\right\}$$

$$= k_0 k_1 \ldots k_{p-1}\lambda_1^{2p}[\mathbf{B} + \mathbf{E}(p)] \qquad (9.2)$$

where $\mathbf{B}$ is independent of p, but the vector $\mathbf{E}(p)$ does depend on p. Let y_{pi}, b_i and $e_i(p)$ denote the ith elements of the vectors $\mathbf{Y}_p$, $\mathbf{B}$ and $\mathbf{E}(p)$ respectively. Then

$$y_{pi} = k_0 k_1 \ldots k_{p-1}\lambda_1^{2p}[b_i + e_i(p)]$$

which means that, if $y_{p-1,i} \neq 0$,

$$\frac{y_{pi}}{y_{p-1,i}} = \frac{k_0 k_1 \ldots k_{p-1}\lambda_1^{2p}[b_i + e_i(p)]}{k_0 k_1 \ldots k_{p-2}\lambda_1^{2(p-1)}[b_i + e_i(p-1)]} = \frac{k_{p-1}\lambda_1^2[b_i + e_i(p)]}{[b_i + e_i(p-1)]}$$

As $p \to \infty$ it is clear that $e_i(p) \to 0$, so that

$$\frac{y_{pi}}{y_{p-1,i}} \to k_{p-1}\lambda_1^2 \qquad (9.3)$$

Since the scalars k_i will be known quantities this gives us the method for finding the latent root of largest modulus, and in general the associated latent vector. The usual choice of k_i is such that the largest component of $k_i\mathbf{Y}_i$ is unity, in which case k_i^{-1} gives the ith estimate of λ_1^2.

The method normally presented forms the sequence

$$\mathbf{Y}_i = k_{i-1}\mathbf{A}\mathbf{Y}_{i-1} \qquad (9.4)$$

and the method given here is really looking at every other vector in this sequence. I believe it has advantages in that it gives faster convergence, does not suffer quite so much from rounding errors and tends to smooth out local instabilities. We look first at a straightforward example.

Example 9.1

$$\mathbf{A} = \begin{pmatrix} 12 & 3 & -11 \\ 1 & 2 & -1 \\ 2 & 3 & -1 \end{pmatrix}, \quad \mathbf{Y}_0 = \begin{pmatrix} 1 \\ 0 \\ 0 \end{pmatrix}, \quad k_0 = 1$$

$$\mathbf{A}^2 = \begin{pmatrix} 125 & 9 & -124 \\ 12 & 4 & -12 \\ 25 & 9 & -24 \end{pmatrix}$$

$$\mathbf{Y}_1 = k_0\mathbf{A}^2\mathbf{Y}_0 = \begin{pmatrix} 125 \\ 12 \\ 25 \end{pmatrix} = 125\begin{pmatrix} 1{\cdot}0000 \\ 0{\cdot}0960 \\ 0{\cdot}2000 \end{pmatrix}, \quad k_1^{-1} = 125$$

$$\mathbf{Y}_2 = k_1\mathbf{A}^2\mathbf{Y}_1 = \begin{pmatrix} 101{\cdot}0640 \\ 9{\cdot}9840 \\ 21{\cdot}0640 \end{pmatrix} = 101{\cdot}064\begin{pmatrix} 1{\cdot}0000 \\ 0{\cdot}0988 \\ 0{\cdot}2084 \end{pmatrix}, \quad k_2^{-1} = 101{\cdot}064$$

$$\mathbf{Y}_3 = k_2\mathbf{A}^2\mathbf{Y}_2 = \begin{pmatrix} 100{\cdot}0476 \\ 9{\cdot}8944 \\ 20{\cdot}8876 \end{pmatrix} = 100{\cdot}0476\begin{pmatrix} 1{\cdot}0000 \\ 0{\cdot}0989 \\ 0{\cdot}2088 \end{pmatrix}, \quad k_3^{-1} = 100{\cdot}0476$$

$$\mathbf{Y}_4 = k_3\mathbf{A}^2\mathbf{Y}_3 = \begin{pmatrix} 99{\cdot}9989 \\ 9{\cdot}8900 \\ 20{\cdot}8789 \end{pmatrix} = 99{\cdot}9989\begin{pmatrix} 1{\cdot}0000 \\ 0{\cdot}0989 \\ 0{\cdot}2088 \end{pmatrix}, \quad k_4^{-1} = 99{\cdot}9989$$

Clearly holding four decimal places we cannot improve the solution further. Now

$$\sqrt{k_4^{-1}} = \pm 9{\cdot}9999$$

and a check shows that the positive sign is to be taken. This agrees very well with the correct root of 10 and the latent vector, which is

$$\begin{pmatrix} 1{\cdot}0000 \\ 0{\cdot}0989 \\ 0{\cdot}2088 \end{pmatrix}$$

correct to four decimal places. The convergence here is good because the other two roots of **A** are $\lambda_2 = 2$ and $\lambda_1 = 1$, so

$$\frac{\lambda_2^2}{\lambda_1^2} = \frac{1}{25} \quad \text{and} \quad \frac{\lambda_3^2}{\lambda_1^2} = \frac{1}{100}$$

and the vector $\mathbf{E}(p)$ in equation (9.2) tends to zero quite quickly. In this example if equation (9.4) is used we find, using the same $\mathbf{Y}_0$,

$$\mathbf{Y}_5 = \begin{pmatrix} 10{\cdot}0040 \\ 0{\cdot}9892 \\ 2{\cdot}0880 \end{pmatrix} = 10{\cdot}004 \begin{pmatrix} 1{\cdot}0000 \\ 0{\cdot}0989 \\ 0{\cdot}2087 \end{pmatrix}, \quad \mathbf{k}_5^{-1} = 10{\cdot}004$$

$$\mathbf{Y}_6 = k_5 \mathbf{A} \mathbf{Y}_5 = \begin{pmatrix} 10{\cdot}0020 \\ 0{\cdot}9891 \\ 2{\cdot}0880 \end{pmatrix} = 10{\cdot}002 \begin{pmatrix} 1{\cdot}0000 \\ 0{\cdot}0989 \\ 0{\cdot}2087 \end{pmatrix}, \quad \mathbf{k}_6^{-1} = 10{\cdot}002$$

and holding four decimal places we can only achieve two-place accuracy in the latent root as opposed to three-place accuracy previously. This is, of course, due to rounding errors, which are not bad in this example because the matrix **A** is well conditioned. In fact, in this example there has been no real gain by using $\mathbf{A}^2$, for we first had to find $\mathbf{A}^2$ and then check using **A** which sign we should select for the root.

Example 9.2

$$\mathbf{A} = \begin{pmatrix} 11 & 2 & -10 \\ -2 & -10 & 2 \\ 1 & 2 & 0 \end{pmatrix}, \quad \mathbf{Y}_0 = \begin{pmatrix} 1 \\ 0 \\ 0 \end{pmatrix}, \quad k_0 = 1$$

$$\mathbf{A}^2 = \begin{pmatrix} 107 & -18 & -106 \\ 0 & 100 & 0 \\ 7 & -18 & -6 \end{pmatrix}$$

$$\mathbf{Y}_1 = k_0 \mathbf{A}^2 \mathbf{Y}_0 = \begin{pmatrix} 107 \\ 0 \\ 7 \end{pmatrix} = 107 \begin{pmatrix} 1{\cdot}0000 \\ 0{\cdot}0000 \\ 0{\cdot}0654 \end{pmatrix}, \quad k_1^{-1} = 107$$

$$\mathbf{Y}_2 = k_1\mathbf{A}^2\mathbf{Y}_1 = \begin{pmatrix} 100{\cdot}0676 \\ 0{\cdot}0000 \\ 6{\cdot}6076 \end{pmatrix} = 100{\cdot}0676\begin{pmatrix} 1{\cdot}0000 \\ 0{\cdot}0000 \\ 0{\cdot}0660 \end{pmatrix}, \quad \mathbf{k}_2^{-1} = 100{\cdot}0676$$

$$\mathbf{Y}_3 = k_2\mathbf{A}^2\mathbf{Y}_2 = \begin{pmatrix} 100{\cdot}0040 \\ 0{\cdot}0000 \\ 6{\cdot}6040 \end{pmatrix} = 100{\cdot}004\begin{pmatrix} 1{\cdot}0000 \\ 0{\cdot}0000 \\ 0{\cdot}0660 \end{pmatrix}, \quad k_3^{-1} = 100{\cdot}0040$$

Again, holding four decimal places, we cannot improve the solution further. Now

$$\sqrt{k_3^{-1}} = \pm 10{\cdot}0002$$

Checking to determine which sign we should select we get

$$\begin{pmatrix} 11 & 2 & -10 \\ -2 & -10 & 2 \\ 1 & 2 & 0 \end{pmatrix}\begin{pmatrix} 1{\cdot}0000 \\ 0{\cdot}0000 \\ 0{\cdot}0660 \end{pmatrix} = \begin{pmatrix} 10{\cdot}3400 \\ -1{\cdot}8680 \\ 1{\cdot}0000 \end{pmatrix} = 10{\cdot}34\begin{pmatrix} 1{\cdot}0000 \\ -0{\cdot}1807 \\ 0{\cdot}0967 \end{pmatrix}$$

and it appears that we have not in fact solved the problem. The reason for this is that **A** actually has one root of 10 and another of -10, and the vector

$$\begin{pmatrix} 1{\cdot}0000 \\ 0{\cdot}0000 \\ 0{\cdot}0660 \end{pmatrix}$$

is not a latent vector of **A**, but a linear combination of the two latent vectors corresponding to the latent roots 10 and -10.† In a case such as this it is generally of advantage to use $\mathbf{A}^2$.

Example 9.3

Here we take the matrix of example 9.1 but start with a different vector.

$$\mathbf{A}^2 = \begin{pmatrix} 125 & 9 & -124 \\ 12 & 4 & -12 \\ 25 & 9 & -24 \end{pmatrix}, \quad \mathbf{Y}_0 = \begin{pmatrix} 1 \\ 1 \\ 1 \end{pmatrix}, \quad k_0 = 1$$

$$\mathbf{Y}_1 = k_0\mathbf{A}^2\mathbf{Y}_0 = \begin{pmatrix} 10 \\ 4 \\ 10 \end{pmatrix} = 10\begin{pmatrix} 1{\cdot}00 \\ 0{\cdot}40 \\ 1{\cdot}00 \end{pmatrix}, \quad k_1^{-1} = 10$$

† We note that if **X** is a latent vector of **A** then by theorem 1.6 it is a latent vector of $\mathbf{A}^2$, but a latent vector of $\mathbf{A}^2$ is not necessarily a latent vector of **A**.

$$\mathbf{Y}_2 = k_1 \mathbf{A}^2 \mathbf{Y}_1 = \begin{pmatrix} 4{\cdot}60 \\ 1{\cdot}60 \\ 4{\cdot}60 \end{pmatrix} = 4{\cdot}60 \begin{pmatrix} 1{\cdot}00 \\ 0{\cdot}35 \\ 1{\cdot}00 \end{pmatrix}, \quad k_2^{-1} = 4{\cdot}60$$

$$\mathbf{Y}_3 = k_2 \mathbf{A}^2 \mathbf{Y}_2 = \begin{pmatrix} 4{\cdot}15 \\ 1{\cdot}40 \\ 4{\cdot}15 \end{pmatrix} = 4{\cdot}15 \begin{pmatrix} 1{\cdot}00 \\ 0{\cdot}34 \\ 1{\cdot}00 \end{pmatrix}, \quad k_3^{-1} = 4{\cdot}15$$

$$\mathbf{Y}_4 = k_3 \mathbf{A}^2 \mathbf{Y}_3 = \begin{pmatrix} 4{\cdot}06 \\ 1{\cdot}36 \\ 4{\cdot}06 \end{pmatrix} = 4{\cdot}06 \begin{pmatrix} 1{\cdot}00 \\ 0{\cdot}33 \\ 1{\cdot}00 \end{pmatrix}, \quad k_4^{-1} = 4{\cdot}06$$

$$\mathbf{Y}_5 = k_4 \mathbf{A}^2 \mathbf{Y}_4 = \begin{pmatrix} 3{\cdot}97 \\ 1{\cdot}32 \\ 3{\cdot}97 \end{pmatrix} = 3{\cdot}97 \begin{pmatrix} 1{\cdot}00 \\ 0{\cdot}33 \\ 1{\cdot}00 \end{pmatrix}, \quad k_5^{-1} = 3{\cdot}97$$

We saw in example 9·1 that the latent root of largest modulus is 10, but obviously we have not converged to this root here. To find the reason for this we have to look back to our original assumptions. In equation (9.1) we assumed that $a_1 \neq 0$. Now the latent vectors corresponding to the two latent roots 2 and 1 of **A** are

$$\mathbf{X}_2 = \begin{pmatrix} 1 \\ \frac{1}{3} \\ 1 \end{pmatrix}, \quad \mathbf{X}_3 = \begin{pmatrix} 1 \\ 0 \\ 1 \end{pmatrix}$$

and it is easily seen that here

$$\mathbf{Y}_0 = 3\mathbf{X}_2 - 2\mathbf{X}_3$$

Hence $a_1 = 0$ and $\lambda_2 = 2$ takes up the role of the dominant latent root. Often rounding errors will cause a small component of $\mathbf{X}_1$ to be introduced, and then after an unstable start convergence to λ_1 will eventually take place. It is not usually pointed out, however, that examples, such as this one, exist where this cannot happen. With $\mathbf{Y}_0$ as the starting vector it is clear that $\mathbf{Y}_i$ is always of the form

$$\mathbf{Y}_i = \begin{pmatrix} x_1 \\ x_2 \\ x_1 \end{pmatrix}$$

which can be expressed as

$$\mathbf{Y}_i = 3x_2\mathbf{X}_2 + (x_1 - 3x_2)\mathbf{X}_3$$

This means that no component of $\mathbf{X}_1$ can ever be introduced. Admittedly this case is rare, but its possibility means that it is generally worth performing a few iterations with another starting vector just to be on the safe side. It is interesting to note that we can have this condition and still converge to the correct latent root, as the next example shows.

Example 9.4

Here we take the matrix $\mathbf{A}$ of example 3.1 which has roots

$$\lambda_1 = \lambda_2 = 3, \quad \lambda_3 = 1$$

$$\mathbf{A} = \begin{pmatrix} 0 & -2 & 5 \\ -7 & 1 & 9 \\ -1 & -2 & 6 \end{pmatrix}, \quad \mathbf{Y}_0 = \begin{pmatrix} 1 \\ 0 \\ 0 \end{pmatrix}, \quad k_0 = 1$$

$$\mathbf{A}^2 = \begin{pmatrix} 9 & -12 & 12 \\ -16 & -3 & 28 \\ 8 & -12 & 13 \end{pmatrix}$$

$$\mathbf{Y}_1 = k_0\mathbf{A}^2\mathbf{Y}_0 = \begin{pmatrix} 9 \\ -16 \\ 8 \end{pmatrix} = -16\begin{pmatrix} -0{\cdot}56 \\ 1{\cdot}00 \\ -0{\cdot}50 \end{pmatrix}, \quad k_1^{-1} = -16$$

$$\mathbf{Y}_2 = k_1\mathbf{A}^2\mathbf{Y}_1 = \begin{pmatrix} -23{\cdot}04 \\ -8{\cdot}04 \\ -22{\cdot}98 \end{pmatrix} = -23{\cdot}04\begin{pmatrix} 1{\cdot}00 \\ 0{\cdot}35 \\ 1{\cdot}00 \end{pmatrix}, \quad k_2^{-1} = -23{\cdot}04$$

$$\mathbf{Y}_3 = k_2\mathbf{A}^2\mathbf{Y}_2 = \begin{pmatrix} 16{\cdot}80 \\ 10{\cdot}95 \\ 16{\cdot}80 \end{pmatrix} = 16{\cdot}80\begin{pmatrix} 1{\cdot}00 \\ 0{\cdot}65 \\ 1{\cdot}00 \end{pmatrix}, \quad k_3^{-1} = 16{\cdot}80$$

$$\mathbf{Y}_4 = k_3\mathbf{A}^2\mathbf{Y}_3 = \begin{pmatrix} 13{\cdot}20 \\ 10{\cdot}05 \\ 13{\cdot}20 \end{pmatrix} = 13{\cdot}20\begin{pmatrix} 1{\cdot}00 \\ 0{\cdot}76 \\ 1{\cdot}00 \end{pmatrix}, \quad k_4^{-1} = 13{\cdot}20$$

$$\mathbf{Y}_5 = k_4\,\mathbf{A}^2\,\mathbf{Y}_4 = \begin{pmatrix} 11{\cdot}88 \\ 9{\cdot}72 \\ 11{\cdot}88 \end{pmatrix} = 11{\cdot}88\begin{pmatrix} 1{\cdot}00 \\ 0{\cdot}82 \\ 1{\cdot}00 \end{pmatrix}, \quad k_5^{-1} = 11{\cdot}88$$

$$\mathbf{Y}_6 = k_5\,\mathbf{A}^2\,\mathbf{Y}_5 = \begin{pmatrix} 11{\cdot}16 \\ 9{\cdot}54 \\ 11{\cdot}16 \end{pmatrix} = 11{\cdot}16\begin{pmatrix} 1{\cdot}00 \\ 0{\cdot}85 \\ 1{\cdot}00 \end{pmatrix}, \quad k_6^{-1} = 11{\cdot}16$$

$$\mathbf{Y}_7 = k_6\,\mathbf{A}^2\,\mathbf{Y}_6 = \begin{pmatrix} 10{\cdot}80 \\ 9.45 \\ 10{\cdot}80 \end{pmatrix} = 10{\cdot}80\begin{pmatrix} 1{\cdot}00 \\ 0{\cdot}88 \\ 1{\cdot}00 \end{pmatrix}, \quad k_7^{-1} = 10{\cdot}80$$

$$\mathbf{Y}_8 = k_7\,\mathbf{A}^2\,\mathbf{Y}_7 = \begin{pmatrix} 10{\cdot}44 \\ 9{\cdot}36 \\ 10{\cdot}44 \end{pmatrix} = 10{\cdot}44\begin{pmatrix} 1{\cdot}00 \\ 0{\cdot}90 \\ 1{\cdot}00 \end{pmatrix}, \quad k_8^{-1} = 10{\cdot}44$$

$$\mathbf{Y}_9 = k_8\,\mathbf{A}^2\,\mathbf{Y}_8 = \begin{pmatrix} 10{\cdot}20 \\ 9{\cdot}30 \\ 10{\cdot}20 \end{pmatrix} = 10{\cdot}20\begin{pmatrix} 1{\cdot}00 \\ 0{\cdot}91 \\ 1{\cdot}00 \end{pmatrix}, \quad k_9^{-1} = 10{\cdot}20$$

$$\mathbf{Y}_{10} = k_9\,\mathbf{A}^2\,\mathbf{Y}_9 = \begin{pmatrix} 10{\cdot}08 \\ 9{\cdot}27 \\ 10{\cdot}08 \end{pmatrix} = 10{\cdot}08\begin{pmatrix} 1{\cdot}00 \\ 0{\cdot}92 \\ 1{\cdot}00 \end{pmatrix}, \quad k_{10}^{-1} = 10{\cdot}08$$

$$\mathbf{Y}_{11} = k_{10}\,\mathbf{A}^2\,\mathbf{Y}_{10} = \begin{pmatrix} 9{\cdot}96 \\ 9{\cdot}24 \\ 9{\cdot}96 \end{pmatrix} = 9{\cdot}96\begin{pmatrix} 1{\cdot}00 \\ 0{\cdot}93 \\ 1{\cdot}00 \end{pmatrix}, \quad k_{11}^{-1} = 9{\cdot}96$$

$$\mathbf{Y}_{12} = k_{11}\,\mathbf{A}^2\,\mathbf{Y}_{11} = \begin{pmatrix} 9{\cdot}84 \\ 9{\cdot}21 \\ 9{\cdot}84 \end{pmatrix} = 9{\cdot}84\begin{pmatrix} 1{\cdot}00 \\ 0{\cdot}94 \\ 1{\cdot}00 \end{pmatrix}, \quad k_{12}^{-1} = 9{\cdot}84$$

$$\mathbf{Y}_{13} = k_{12}\,\mathbf{A}^2\,\mathbf{Y}_{12} = \begin{pmatrix} 9{\cdot}72 \\ 9{\cdot}18 \\ 9{\cdot}72 \end{pmatrix} = 9{\cdot}72\begin{pmatrix} 1{\cdot}00 \\ 0{\cdot}94 \\ 1{\cdot}00 \end{pmatrix}, \quad k_{13}^{-1} = 9{\cdot}72$$

and holding two decimal places this is as far as we can go. Now

$$\sqrt{k_{13}^{-1}} = 3{\cdot}12$$

and we have not even achieved one-decimal place accuracy. Also, remembering that $\lambda_1 = \lambda_2$ we have

$$\frac{\lambda_3^2}{\lambda_1^2} = \frac{1}{9}$$

and $(\frac{1}{9})^{13} \simeq 0{\cdot}3 \times 10^{-10}$, so we would hardly expect such slow convergence. The inaccuracy in the solution is due to rounding errors, this matrix being particularly badly conditioned. If we use the sequence of vectors defined by equation (9.4) and hold two decimal places, the convergence stops when $\lambda_1 = 3{\cdot}20$ and the latent vector is

$$\begin{pmatrix} 1{\cdot}00 \\ 0{\cdot}90 \\ 1{\cdot}00 \end{pmatrix}$$

The correct latent vector is given by $\mathbf{X}_1^T = (\,1 \quad 1 \quad 1\,)$. We can see that in situations such as these there is an advantage in working with $\mathbf{A}^2$. The reason for the poor convergence is that the matrix $\mathbf{A}$ has only two latent vectors, these being†

$$\mathbf{X}_1 = \begin{pmatrix} 1 \\ 1 \\ 1 \end{pmatrix}, \quad \mathbf{X}_2 = \begin{pmatrix} 1 \\ \frac{13}{9} \\ \frac{7}{9} \end{pmatrix}$$

$\mathbf{Y}_0$ cannot be expressed as a linear combination of $\mathbf{X}_1$ and $\mathbf{X}_2$, as is required by equation (9.1). Here we have the remarkable situation that, even if we start with a vector which is a linear combination of $\mathbf{X}_1$ and $\mathbf{X}_2$, rounding errors will tend to destroy this linear dependence.

Example 9.5

$$\mathbf{A} = \begin{pmatrix} 4 & 1 & -3 \\ -5 & -2 & 5 \\ 2 & -3 & -1 \end{pmatrix}, \quad \mathbf{Y}_0 = \begin{pmatrix} 1 \\ 0 \\ 0 \end{pmatrix}, \quad k_0 = 1$$

$$\mathbf{A}^2 = \begin{pmatrix} 5 & 11 & -4 \\ 0 & -16 & 0 \\ 21 & 11 & -20 \end{pmatrix}$$

† See example 3.4.

$$\mathbf{Y}_1 = k_0\mathbf{A}^2\mathbf{Y}_0 = \begin{pmatrix} 5 \\ 0 \\ 21 \end{pmatrix} = 21\begin{pmatrix} 0{\cdot}24 \\ 0{\cdot}00 \\ 1{\cdot}00 \end{pmatrix}, \quad k_1^{-1} = 21$$

$$\mathbf{Y}_2 = k_1\mathbf{A}^2\mathbf{Y}_1 = \begin{pmatrix} -2{\cdot}80 \\ 0{\cdot}00 \\ -14{\cdot}96 \end{pmatrix} = -14{\cdot}96\begin{pmatrix} 0{\cdot}19 \\ 0{\cdot}00 \\ 1{\cdot}00 \end{pmatrix}, \quad k_2^{-1} = -14{\cdot}96$$

$$\mathbf{Y}_3 = k_2\mathbf{A}^2\mathbf{Y}_2 = \begin{pmatrix} -3{\cdot}05 \\ 0{\cdot}00 \\ -16{\cdot}01 \end{pmatrix} = -16{\cdot}01\begin{pmatrix} 0{\cdot}19 \\ 0{\cdot}00 \\ 1{\cdot}00 \end{pmatrix}, \quad k_3^{-1} = -16{\cdot}01$$

and holding two decimal places this is as far as we can go. This is an interesting example for we have converged to a negative value, the reason being that the required root of **A** is imaginary. Now

$$\sqrt{k_3^{-1}} = \pm 4{\cdot}00i$$

which corresponds exactly to the two roots of largest modulus of **A**. Here, as in example 9.2, the vector

$$\begin{pmatrix} 0{\cdot}19 \\ 0{\cdot}00 \\ 1{\cdot}00 \end{pmatrix}$$

is a linear combination of the latent vectors of **A**. Clearly in a case such as this it is of advantage to use $\mathbf{A}^2$.

Example 9.6

$$\mathbf{A} = \begin{pmatrix} 4 & 1 & 16 \\ 2 & 20 & -3 \\ 3 & 1 & 17 \end{pmatrix}, \quad \mathbf{Y}_0 = \begin{pmatrix} 1 \\ 0 \\ 0 \end{pmatrix}, \quad k_0 = 1$$

$$\mathbf{A}^2 = \begin{pmatrix} 66 & 40 & 333 \\ 39 & 399 & -79 \\ 65 & 40 & 334 \end{pmatrix}$$

$$\mathbf{Y}_1 = k_0\mathbf{A}^2\mathbf{Y}_0 = \begin{pmatrix} 66 \\ 39 \\ 65 \end{pmatrix} = 66\begin{pmatrix} 1{\cdot}00 \\ 0{\cdot}59 \\ 0{\cdot}98 \end{pmatrix}, \quad k_1^{-1} = 66$$

$$\mathbf{Y}_2 = k_1\mathbf{A}^2\mathbf{Y}_1 = \begin{pmatrix} 415\cdot94 \\ 196\cdot99 \\ 415\cdot92 \end{pmatrix} = 415\cdot94\begin{pmatrix} 1\cdot00 \\ 0\cdot47 \\ 1\cdot00 \end{pmatrix}, \quad k_2^{-1} = 415\cdot94$$

$$\mathbf{Y}_3 = k_2\mathbf{A}^2\mathbf{Y}_2 = \begin{pmatrix} 417\cdot80 \\ 147\cdot53 \\ 417\cdot80 \end{pmatrix} = 417\cdot80\begin{pmatrix} 1\cdot00 \\ 0\cdot35 \\ 1\cdot00 \end{pmatrix}, \quad k_3^{-1} = 417\cdot80$$

$$\mathbf{Y}_4 = k_3\mathbf{A}^2\mathbf{Y}_3 = \begin{pmatrix} 413\cdot00 \\ 99\cdot65 \\ 413\cdot00 \end{pmatrix} = 413\cdot00\begin{pmatrix} 1\cdot00 \\ 0\cdot24 \\ 1\cdot00 \end{pmatrix}, \quad k_4^{-1} = 413\cdot00$$

$$\mathbf{Y}_5 = k_4\mathbf{A}^2\mathbf{Y}_4 = \begin{pmatrix} 408\cdot60 \\ 55\cdot76 \\ 408\cdot60 \end{pmatrix} = 408\cdot60\begin{pmatrix} 1\cdot00 \\ 0\cdot14 \\ 1\cdot00 \end{pmatrix}, \quad k_5^{-1} = 408\cdot6$$

$$\mathbf{Y}_6 = k_5\mathbf{A}^2\mathbf{Y}_5 = \begin{pmatrix} 404\cdot60 \\ 15\cdot86 \\ 404\cdot60 \end{pmatrix} = 404\cdot60\begin{pmatrix} 1\cdot00 \\ 0\cdot04 \\ 1\cdot00 \end{pmatrix}, \quad k_6^{-1} = 404\cdot60$$

$$\mathbf{Y}_7 = k_6\mathbf{A}^2\mathbf{Y}_6 = \begin{pmatrix} 400\cdot60 \\ -24\cdot04 \\ 400\cdot60 \end{pmatrix} = 400\cdot60\begin{pmatrix} 1\cdot00 \\ -0\cdot06 \\ 1\cdot00 \end{pmatrix}, \quad k_7^{-1} = 400\cdot60$$

It appears here that no convergence is occurring. In such an instance we suspect the matrix of having complex roots. In the next section we shall see how these may be found without using complex arithmetic.

9.2 Finding Complex Roots

Once we have decided we are looking for complex roots it is better to revert to using **A**, instead of $\mathbf{A}^2$, and the sequence defined by equation (9.4), because, as will be seen, the roots are then easier to find.

Let the roots of largest modulus of **A** be λ_1 and λ_2, and let

$$\lambda_1 = \alpha + i\beta \quad \text{so that} \quad \lambda_2 = \alpha - i\beta$$

Also, let the corresponding latent vectors be $\mathbf{X}_1$ and $\mathbf{X}_2$ where these are also complex conjugates.†

† See theorem 1.19.

Now, remembering that we are using the sequence of vectors of equation (9.4), it is easily shown that the equation equivalent to (9.2) becomes

$$Y_p = k_0 k_1 \dots k_{p-1}[a_1 \lambda_1^p \mathbf{X}_1 + a_2 \lambda_2^p \mathbf{X}_2 + E(\mathbf{p})]$$

where a_1 and a_2 are complex conjugates. Obviously λ_1 and λ_2 both satisfy the same quadratic equation, say

$$\lambda^2 + b_1 \lambda + b_2 = 0$$

Using this fact we find that

$$b_2 \mathbf{Y}_p + \frac{b_1 \mathbf{Y}_{p+1}}{k_p} + \frac{\mathbf{Y}_{p+2}}{k_p k_{p+1}} = k_0 k_1 \dots k_{p-1}[b_2 \mathbf{E}(p) + b_1 \mathbf{E}(p+1) + \mathbf{E}(p+2)]$$

Hence as $p \to \infty$,

$$b_2 \mathbf{Y}_p + \frac{b_1 \mathbf{Y}_{p+1}}{k_p} + \frac{\mathbf{Y}_{p+2}}{k_p k_{p+1}} \to 0 \tag{9.5}$$

which enables us to find b_1 and b_2.† Furthermore, if we put

$$a_1 \lambda_1^p \mathbf{X}_1 = \mathbf{Z}_1 + i\mathbf{Z}_2$$

then

$$a_2 \lambda_2^p \mathbf{X}_2 = \mathbf{Z}_1 - i\mathbf{Z}_2$$

and clearly as $p \to \infty$,

$$\mathbf{Y}_p \to 2k_0 k_1 \dots k_{p-1} \mathbf{Z}_1 \tag{9.6}$$

Also,

$$\begin{aligned} \mathbf{Y}_{p+1} &= k_0 k_1 \dots k_p[a_1 \lambda_1^{p+1} \mathbf{X}_1 + a_2 \lambda_2^{p+1} \mathbf{X}_2 + \mathbf{E}(p+1)] \\ &= k_0 k_1 \dots k_p[\lambda_1(\mathbf{Z}_1 + i\mathbf{Z}_2) + \lambda_2(\mathbf{Z}_1 - i\mathbf{Z}_2) + \mathbf{E}(p+1)] \\ &= k_0 k_1 \dots k_p[2\alpha \mathbf{Z}_1 - 2\beta \mathbf{Z}_2 + \mathbf{E}(p+1)] \end{aligned} \tag{9.7}$$

Equations (9.6) and (9.7) allow us to find $\mathbf{Z}_1$ and $\mathbf{Z}_2$, and hence the latent vectors associated with λ_1 and λ_2.

Example 9.7

We take here the matrix of example (9.6) which we already suspect of having complex roots, and use as our starting vector the vector $k_7 \mathbf{Y}_7$ of that example.

† If we were still working with $\mathbf{A}^2$ the roots of the quadratic would be λ_1^2 and λ_2^2 and some complex arithmetic would be required to find λ_1 and λ_2.

$$\mathbf{A} = \begin{pmatrix} 4 & 1 & 16 \\ 2 & 20 & -3 \\ 3 & 1 & 17 \end{pmatrix}, \quad \mathbf{Y}_0 = \begin{pmatrix} 1\cdot00 \\ -0\cdot06 \\ 1\cdot00 \end{pmatrix}, \quad k_0 = 1$$

$$\mathbf{Y}_1 = k_0\mathbf{A}\mathbf{Y}_0 = \begin{pmatrix} 19\cdot94 \\ -2\cdot20 \\ 19\cdot94 \end{pmatrix}, \quad k_1 = 1$$

$$\mathbf{Y}_2 = k_1\mathbf{A}\mathbf{Y}_1 = \begin{pmatrix} 396\cdot60 \\ -63\cdot94 \\ 396\cdot60 \end{pmatrix}$$

Now, putting

$$b_2\mathbf{Y}_0 + \frac{b_1\mathbf{Y}_1}{k_0} + \frac{\mathbf{Y}_2}{k_0 k_1} = 0$$

we get

$$b_2\begin{pmatrix} 1\cdot00 \\ -0\cdot06 \\ 1\cdot00 \end{pmatrix} + b_1\begin{pmatrix} 19\cdot94 \\ -2\cdot20 \\ 19\cdot94 \end{pmatrix} = \begin{pmatrix} -396\cdot60 \\ 63\cdot94 \\ -396\cdot60 \end{pmatrix}$$

Solving for b_1 and b_2 we find†

$$b_1 = -40 \quad \text{and} \quad b_2 = 401$$

By solving the quadratic equation $\lambda^2 - 40\lambda + 401 = 0$ we obtain

$$\lambda_1 = 20 + i \quad \text{and} \quad \lambda_2 = 20 - i$$

which are the exact latent roots of **A**.

From equation (9.6) we put

$$\mathbf{Y}_1 = 2k_0\mathbf{Z}_1$$

so that

$$\mathbf{Z}_1 = \begin{pmatrix} 9\cdot97 \\ -1\cdot10 \\ 9\cdot97 \end{pmatrix}$$

† Note that here if we solve the first two equations the third one is automatically satisfied. This will not generally be so, and we can perhaps use the third to determine whether or not we need more iterates.

and from equation (9.7) we put

$$\mathbf{Y}_2 = k_0 k_1[2\alpha\mathbf{Z}_1 - 2\beta\mathbf{Z}_2]$$

so that

$$\mathbf{Z}_2 = \frac{1}{2\beta}\left[2\alpha\mathbf{Z}_1 - \frac{\mathbf{Y}_2}{k_0 k_1}\right]$$

$$= \frac{1}{2}\left\{40\begin{pmatrix} 9\cdot97 \\ -1\cdot10 \\ 9\cdot97 \end{pmatrix} - \begin{pmatrix} 396\cdot60 \\ -63\cdot94 \\ 396\cdot60 \end{pmatrix}\right\} = \begin{pmatrix} 1\cdot10 \\ 9\cdot97 \\ 1\cdot10 \end{pmatrix}$$

Hence the latent vectors are given by

$$\mathbf{X}_1 = \begin{pmatrix} 9\cdot97+1\cdot10i \\ -1\cdot10+9\cdot97i \\ 9\cdot97+1\cdot10i \end{pmatrix} \quad \text{and} \quad \mathbf{X}_2 = \begin{pmatrix} 9\cdot97-1\cdot10i \\ -1\cdot10-9\cdot97i \\ 9\cdot97-1\cdot10i \end{pmatrix}$$

which correspond to the exact latent vectors of **A**.

Although we reverted back to using **A** instead of $\mathbf{A}^2$, our earlier work was not wasted, for we only required two further iterations. This example has given particularly good results. It should be pointed out that this does not necessarily follow because the quadratic equation

$$\lambda^2 + b_1\lambda + b_2 = 0$$

is often ill-conditioned, especially if β (the imaginary part of λ_1) is small.†

9.3 Improving Convergence

In §9.1 we performed iteration using the simple polynomial in $\mathbf{A}, \mathbf{A}^2$. We can, of course, use any polynomial in **A** to iterate with, because, from theorem 1.13, if λ is a latent root of **A**, then $f(\lambda)$ is a latent root of $f(\mathbf{A})$. Naturally there is not much point in using a polynomial for which it is laborious to find the roots. A polynomial which is often of use in improving convergence is one of the form $\mathbf{A}^2 - d\mathbf{I}$.

Suppose that the latent roots of $\mathbf{A}^2$ are real and that

$$\lambda_1^2 > \lambda_2^2 \geqslant \lambda_3^2 \geqslant \ldots \geqslant \lambda_{n-1}^2 > \lambda_n^2$$

Then for any choice of d, either $\lambda_1^2 - d$ or $\lambda_n^2 - d$ is the dominant latent root of $\mathbf{A}^2 - d\mathbf{I}$. The best convergence to $\lambda_1^2 - d$ will be obtained when

$$d = \tfrac{1}{2}(\lambda_2^2 + \lambda_n^2)$$

† See reference 7, p. 580.

and to $\lambda_n^2 - d$

$$d = \tfrac{1}{2}(\lambda_1^2 + \lambda_{n-1}^2)$$

for it is in these cases that the vector $\mathbf{E}(p)$ of equation (9.2) goes to zero most rapidly.

Example 9.8

Here we take the matrix $\mathbf{A}$ of example 9.1 and, being wise after the event, we take $d = \tfrac{1}{2}(4+1) = 2{\cdot}5$ so that

$$\mathbf{A}^2 - d\mathbf{I} = \begin{pmatrix} 122{\cdot}5 & 9{\cdot}0 & -124{\cdot}0 \\ 12{\cdot}0 & 1{\cdot}5 & -12{\cdot}0 \\ 25{\cdot}0 & 9{\cdot}0 & -26{\cdot}5 \end{pmatrix}$$

Using $\mathbf{Y}_0$ and k_0 as in example 9·1 we get

$$\mathbf{Y}_1 = k_0(\mathbf{A}^2 - d\mathbf{I})\,\mathbf{Y}_0 = \begin{pmatrix} 122{\cdot}5 \\ 12{\cdot}0 \\ 25{\cdot}0 \end{pmatrix} = 122{\cdot}5\begin{pmatrix} 1{\cdot}0000 \\ 0{\cdot}0980 \\ 0{\cdot}2041 \end{pmatrix}, \quad k_1^{-1} = 122{\cdot}5$$

$$\mathbf{Y}_2 = k_1(\mathbf{A}^2 - d\mathbf{I})\,\mathbf{Y}_1 = \begin{pmatrix} 98{\cdot}0736 \\ 9{\cdot}6978 \\ 20{\cdot}4734 \end{pmatrix} = 98{\cdot}0736\begin{pmatrix} 1{\cdot}0000 \\ 0{\cdot}0989 \\ 0{\cdot}2088 \end{pmatrix}, \quad k_2^{-1} = 98{\cdot}0736$$

$$\mathbf{Y}_3 = k_2(\mathbf{A}^2 - d\mathbf{I})\,\mathbf{Y}_2 = \begin{pmatrix} 97{\cdot}4989 \\ 9{\cdot}6428 \\ 20{\cdot}3569 \end{pmatrix} = 97{\cdot}4989\begin{pmatrix} 1{\cdot}0000 \\ 0{\cdot}0989 \\ 0{\cdot}2088 \end{pmatrix}, \quad k_3^{-1} = 97{\cdot}4989$$

and we have now converged to the solution. This gives

$$\lambda_1^2 = 97{\cdot}4989 + d = 99{\cdot}9989$$

We can see that we have got the same solution as in example 9.1, but we have reached it in three iterations instead of four. Whereas

$$\frac{\lambda_2^2}{\lambda_1^2} = \frac{1}{25} \quad \text{and} \quad \frac{\lambda_3^2}{\lambda_1^2} = \frac{1}{100}$$

we now have

$$\frac{\lambda_2^2 - d}{\lambda_1^2 - d} = -\frac{\lambda_3^2 - d}{\lambda_1^2 - d} = \frac{3}{197}$$

so that $\mathbf{E}(p)$ tends to zero somewhat faster.

This device is extremely useful if we have some knowledge as to the dominant and subdominant latent roots, or if convergence is slow because of close roots. In the latter case the rate of convergence will give us an indication as to what value of d should be selected, and if the roots of $\mathbf{A}$ are real, it is especially useful working with $\mathbf{A}^2$ because we know that its roots will be positive. A full discussion of this type of shift of origin is given by Wilkinson,† as well as an acceleration technique due to Aitken.

9.4 Inverse Iteration

Another useful possibility is to iterate with the polynomials $(\mathbf{A}^2 - d\mathbf{I})^{-1}$ or $(\mathbf{A} - d\mathbf{I})^{-1}$. The second case was the one recommended for finding the latent vector of a tridiagonal matrix. Rather than computing the inverse itself we solve at each stage the set of equations

$$(\mathbf{A}^2 - d\mathbf{I})\,\mathbf{Y}_{i+1} = k_i\,\mathbf{Y}_i$$

in order to find $\mathbf{Y}_{i+1}$. It is interesting to notice that in this method, with a suitable choice of d, we can converge to any desired root. It is usual to use a method such as triangular decomposition to solve the equations, altering only the right-hand side at each stage. This method is analysed fully by Wilkinson.‡

Example 9.9

Here we take the matrix $\mathbf{A}$ of example 9.2. We have already determined that $\lambda_1^2 = \lambda_2^2 = 100$ and the rate of convergence suggests that the remaining root is somewhat less than two. Taking $d = 2$ we get

$$\mathbf{A}^2 - d\mathbf{I} = \begin{pmatrix} 105 & -18 & -106 \\ 0 & 98 & 0 \\ 7 & -18 & -8 \end{pmatrix}$$

and taking

$$\mathbf{Y}_0 = \begin{pmatrix} 1 \\ 0 \\ 0 \end{pmatrix}, \quad k_0 = 1$$

then $(\mathbf{A}^2 - 2\mathbf{I})\,\mathbf{Y}_1 = k_0\,\mathbf{Y}_0$ yields the equations

$$\begin{aligned} 105y_1 - 18y_2 - 106y_3 &= 1 \\ 98y_2 \qquad\qquad &= 0 \\ 7y_1 - 18y_2 - 8y_3 &= 0 \end{aligned}$$

† See reference 7, pp. 572–584.
‡ See reference 7, pp. 619–626.

which have the solution

$$y_1 = 0{\cdot}0816, \quad y_2 = 0{\cdot}0000, \quad y_3 = 0{\cdot}0714$$

so that

$$\mathbf{Y}_1 = \begin{pmatrix} 0{\cdot}0816 \\ 0{\cdot}0000 \\ 0{\cdot}0714 \end{pmatrix} = 0{\cdot}0816 \begin{pmatrix} 1{\cdot}0000 \\ 0{\cdot}0000 \\ 0{\cdot}8750 \end{pmatrix}, \quad k_1^{-1} = 0{\cdot}0816$$

Then $(\mathbf{A}^2 - 2\mathbf{I})\,\mathbf{Y}_2 = k_1\,\mathbf{Y}_1$ yields the equations

$$\begin{aligned} 105y_1 - 18y_2 - 106y_3 &= 1{\cdot}0000 \\ 98y_2 \qquad\qquad &= 0{\cdot}0000 \\ 7y_1 - 18y_2 - 8y_3 &= 0{\cdot}8750 \end{aligned}$$

which have the solution

$$y_1 = -0{\cdot}8648, \quad y_2 = 0{\cdot}0000, \quad y_3 = -0{\cdot}8661$$

so that

$$\mathbf{Y}_2 = \begin{pmatrix} -0{\cdot}8648 \\ 0{\cdot}0000 \\ -0{\cdot}8661 \end{pmatrix} = -0{\cdot}8661 \begin{pmatrix} 0{\cdot}9985 \\ 0{\cdot}0000 \\ 1{\cdot}0000 \end{pmatrix}, \quad k_2^{-1} = -0{\cdot}8661$$

Then $(\mathbf{A}^2 - 2\mathbf{I})\,\mathbf{Y}_3 = k_2\,\mathbf{Y}_2$ yields the equations

$$\begin{aligned} 105y_1 - 18y_2 - 106y_3 &= 0{\cdot}9985 \\ 98y_2 \qquad\qquad &= 0{\cdot}0000 \\ 7y_1 - 18y_2 - 8y_3 &= 1{\cdot}0000 \end{aligned}$$

which have the solution

$$y_1 = -1{\cdot}0001, \quad y_2 = 0{\cdot}0000, \quad y_3 = -1{\cdot}0001$$

so that

$$\mathbf{Y}_3 = \begin{pmatrix} -1{\cdot}0001 \\ 0{\cdot}0000 \\ -1{\cdot}0001 \end{pmatrix} = -1{\cdot}0001 \begin{pmatrix} 1{\cdot}0000 \\ 0{\cdot}0000 \\ 1{\cdot}0000 \end{pmatrix}, \quad k_3^{-1} = -1{\cdot}0001$$

Now

$$k_3^{-1} + d = 1{\cdot}0001$$

and $\sqrt{1\cdot0001} = 1\cdot0000$ correct to four decimal places, which corresponds exactly to the correct value of λ_3, as does the latent vector

$$\begin{pmatrix} 1 \\ 0 \\ 1 \end{pmatrix}$$

In this method, as in that of the previous section, we have the possibility of adjusting d as we proceed. This is not to be recommended unless convergence is not taking place or is slow, because this means that we have to alter the left-hand sides of the simultaneous equations. We note that the better the approximation d is to the latent root the more ill-conditioned the equations become, so that care is needed in their solution.

Inverse iteration is one of the most useful of the available iterative methods.

9.5 Matrix Deflation

There are numerous methods available for deflating an $n \times n$ matrix, $\mathbf{A}$, to one of size $(n-1)\times(n-1)$ having $n-1$ of the latent roots of $\mathbf{A}$. This technique is clearly useful in the context of this chapter. I intend to look at one such method here. This is based on the unitary similarity transformations of theorem 1.5.

Suppose that we have determined the latent root λ_1 and the corresponding latent vector $\mathbf{X}_1$ of the matrix $\mathbf{A}$, and that $\mathbf{X}_1$ is normalized so that

$$\mathbf{X}_1^* \mathbf{X}_1 = 1$$

Then, if we take a unitary matrix $\mathbf{C}_1$ having $\mathbf{X}_1$ as its first column, we find as in theorem 1.5 that

$$\mathbf{C}_1^* \mathbf{A}\mathbf{C}_1 = \mathbf{C}_1^{-1} \mathbf{A}\mathbf{C}_1 = \left(\begin{array}{c|cccc} \lambda_1 & c_1 & c_2 & \cdots & c_{n-1} \\ \hline 0 & & & & \\ 0 & & & & \\ \vdots & & \mathbf{A}_{n-1} & & \\ 0 & & & & \end{array}\right) = \mathbf{B}_1$$

where $\mathbf{A}_{n-1}$ is an $(n-1)\times(n-1)$ matrix having $\lambda_2, \lambda_3, \ldots, \lambda_n$ as its latent roots. Once we have found a root of $\mathbf{A}_{n-1}$ we can then repeat the process with $\mathbf{A}_{n-1}$. If we find all the latent roots of $\mathbf{A}$ in this manner we shall eventually produce the triangular matrix of theorem 1.5. Let us write the matrix $\mathbf{B}_1$ as

$$\mathbf{B}_1 = \left(\begin{array}{c|c} \lambda_1 & \mathbf{D}^T \\ \hline 0 & \mathbf{A}_{n-1} \end{array}\right)$$

Then suppose that we have found the latent root λ_2 and the corresponding latent vector $\mathbf{Z}_{n-1}$ of $\mathbf{A}_{n-1}$. We get

$$\mathbf{B}_1 \mathbf{Y}_2 = \left(\begin{array}{c|c} \lambda_1 & \mathbf{D}^T \\ \hline 0 & \mathbf{A}_{n-1} \end{array}\right) \left(\begin{array}{c} z \\ \hline \mathbf{Z}_{n-1} \end{array}\right) = \lambda_2 \left(\begin{array}{c} z \\ \hline \mathbf{Z}_{n-1} \end{array}\right) = \lambda_2 \mathbf{Y}_2$$

which gives

$$\lambda_1 z + \mathbf{D}^T \mathbf{Z}_{n-1} = \lambda_2 z \tag{9.8}$$

and, of course,

$$\mathbf{A}_{n-1} \mathbf{Z}_{n-1} = \lambda_2 \mathbf{Z}_{n-1}$$

Equation (9.8) gives

$$z = (\lambda_2 - \lambda_1)^{-1} \mathbf{D}^T \mathbf{Z}_{n-1} \tag{9.9}$$

which allows us to find the only unknown element of the latent vector $\mathbf{Y}_2$. From theorem 1.2 we have

$$\mathbf{C}_1 \mathbf{Y}_2 = \mathbf{X}_2$$

from which we may find the latent vector $\mathbf{X}_2$ of $\mathbf{A}$.

It remains to be shown how we may find the matrix $\mathbf{C}_1$. We shall consider only the case of a real latent vector, which means that $\mathbf{C}_1$ is orthogonal, in the hope that the extension to the complex case is then clear. It is convenient to choose $\mathbf{C}_1$ so that it is symmetric,† of the form

$$\mathbf{C}_1 = \mathbf{I} - 2\mathbf{Y}\mathbf{Y}^T$$

as in the Householder transformation of § 7.1, which means that

$$\mathbf{C}_1^{-1} = \mathbf{C}_1^T = \mathbf{C}_1$$

If the elements of $\mathbf{Y}$ are $y_1, y_2, \ldots, y_n$, then

$$\mathbf{C}_1 = \mathbf{I} - 2\mathbf{Y}\mathbf{Y}^T = \begin{pmatrix} 1-2y_1^2 & -2y_1 y_2 & \ldots & -2y_1 y_n \\ -2y_1 y_2 & 1-2y_2^2 & \ldots & -2y_2 y_n \\ \vdots & \vdots & & \vdots \\ -2y_1 y_n & -2y_2 y_n & \ldots & 1-2y_n^2 \end{pmatrix}$$

and since we wish the first column of $\mathbf{C}_1$ to be the latent vector $\mathbf{X}_1$ whose components are $x_1, x_2, \ldots, x_n$, we get

$$1-2y_1^2 = x_1, \quad -2y_1 y_2 = x_2, \ldots, \quad -2y_1 y_n = x_n$$

and from these n equations it is easy to determine the values of $y_1, y_2, \ldots, y_n$.

† In the complex case $\mathbf{C}_1$ would be Hermitian.

Subsequent deflations may be similarly performed. Although this method requires more calculations than some of the other deflation methods, it is quite general and is extremely stable.†

Example 9.10

$$\mathbf{A} = \begin{pmatrix} 16 & -8 & 8 \\ -8 & 10 & 20 \\ -20 & 1 & 50 \end{pmatrix}$$

Suppose that iteration has given us the latent root of largest modulus of **A**, which is $\lambda_1 = 48$, and the corresponding latent vector, which is

$$\begin{pmatrix} \frac{1}{8} \\ \frac{1}{2} \\ 1 \end{pmatrix}$$

Then normalizing so that $\mathbf{X}_1^T \mathbf{X}_1 = 1$ we get

$$\mathbf{X}_1 = \begin{pmatrix} \frac{1}{9} \\ \frac{4}{9} \\ \frac{8}{9} \end{pmatrix}$$

This means that

$$1 - 2y_1^2 = \tfrac{1}{9}$$
$$-2y_1 y_2 = \tfrac{4}{9}$$
$$-2y_1 y_3 = \tfrac{8}{9}$$

and hence

$$y_1 = \tfrac{2}{3}, \quad y_2 = -\tfrac{1}{3}, \quad y_3 = -\tfrac{2}{3}$$

This gives

$$\mathbf{C}_1^{-1}\mathbf{A}\mathbf{C}_1 = \mathbf{C}_1\mathbf{A}\mathbf{C}_1$$

$$= \begin{pmatrix} \frac{1}{9} & \frac{4}{9} & \frac{8}{9} \\ \frac{4}{9} & \frac{7}{9} & -\frac{4}{9} \\ \frac{8}{9} & -\frac{4}{9} & \frac{1}{9} \end{pmatrix} \begin{pmatrix} 16 & -8 & 8 \\ -8 & 10 & 20 \\ -20 & 1 & 50 \end{pmatrix} \begin{pmatrix} \frac{1}{9} & \frac{4}{9} & \frac{8}{9} \\ \frac{4}{9} & \frac{7}{9} & -\frac{4}{9} \\ \frac{8}{9} & -\frac{4}{9} & \frac{1}{9} \end{pmatrix}$$

$$= \begin{pmatrix} 48 & -\frac{88}{3} & -\frac{40}{3} \\ 0 & \frac{26}{3} & \frac{20}{3} \\ 0 & -\frac{11}{3} & \frac{58}{3} \end{pmatrix}$$

† See reference 7, p. 594.

We can now find the next latent root and vector using the deflated matrix

$$\mathbf{A}_2 = \begin{pmatrix} \frac{26}{3} & \frac{20}{3} \\ -\frac{11}{3} & \frac{58}{3} \end{pmatrix}$$

This gives $\lambda_2 = 16$ and

$$\mathbf{Z}_2 = \begin{pmatrix} \frac{10}{11} \\ 1 \end{pmatrix}$$

From equation (9.9) we get

$$z = (16-48)^{-1}\begin{pmatrix} -\frac{88}{3} & -\frac{40}{3} \end{pmatrix} \begin{pmatrix} \frac{10}{11} \\ 1 \end{pmatrix} = \begin{pmatrix} \frac{5}{4} \end{pmatrix}$$

so that the latent vector of $\mathbf{A}$ is given by

$$\mathbf{X}_2 = \mathbf{C}_1 \mathbf{Y}_2 = \begin{pmatrix} \frac{1}{9} & \frac{4}{9} & \frac{8}{9} \\ \frac{4}{9} & \frac{7}{9} & -\frac{4}{9} \\ \frac{8}{9} & -\frac{4}{9} & \frac{1}{9} \end{pmatrix} \begin{pmatrix} \frac{5}{4} \\ \frac{10}{11} \\ 1 \end{pmatrix} = \begin{pmatrix} \frac{63}{44} \\ \frac{9}{11} \\ \frac{9}{11} \end{pmatrix} = \frac{63}{44}\begin{pmatrix} 1 \\ \frac{4}{7} \\ \frac{4}{7} \end{pmatrix}$$

We could similarly deflate $\mathbf{A}_2$ to obtain the remaining latent root $\lambda_3 = 12$ and the latent vector,

$$\mathbf{X}_3 = \begin{pmatrix} 1 \\ 1 \\ \frac{1}{2} \end{pmatrix}$$

Deflation is often a very useful technique, especially for large matrices when only a few of the latent roots and vectors are required. This method is also useful in conjuction with § 2.4 in the case of equal roots.

9.6 Further Comments on the Iterative Method

It is clear that there are many circumstances under which an iterative method is extremely useful. The case of complex roots needs especial care because the quadratic equation can be ill-conditioned, but since we are able to find two roots simultaneously it is not really surprising that we are likely to need extra precision. The case that gives real difficulty is that of a matrix not possessing n linearly independent latent vectors.† Methods for dealing with this are discussed by Wilkinson.‡

Although in nearly every instance we have taken k_i^{-1} to be the element of largest modulus in the vector $\mathbf{Y}_i$, it is occasionally worth taking $k_i = 1$

† See example 9.4.
‡ See reference 7, Chapter 9, §§ 16, 32, 41 and 53.

for certain i, especially for the first few iterates, to iron out local instabilities or to avoid rounding errors due to division of the elements in $\mathbf{Y}_i$.

It is hoped that a reasonable case has been made for the use of $\mathbf{A}^2$ in preference to $\mathbf{A}$. Nevertheless, it is often quite adequate to use $\mathbf{A}$, especially if prior knowledge shows that no advantage is to be gained by using $\mathbf{A}^2$. We note that using $\mathbf{A}^2$ approximately halves the number of iterations required, but we have roughly n^3 extra calculations to make in finding $\mathbf{A}^2$.

An interesting discussion of various iterative methods is also given by Householder.†

9.7 Exercises

9.1. Find the latent roots of largest modulus and corresponding latent vectors of the following matrices:

$$\text{(i)}\quad \mathbf{A} = \begin{pmatrix} 3 & 4 & -4 \\ 2 & 1 & -2 \\ -6 & 4 & 5 \end{pmatrix}, \qquad \text{(ii)}\quad \mathbf{A} = \begin{pmatrix} 5 & 2 & -20 \\ 3 & 1 & -3 \\ -10 & 2 & -5 \end{pmatrix}$$

$$\text{(iii)}\quad \mathbf{A} = \begin{pmatrix} 13 & 7 & -12 \\ -5 & 10 & 5 \\ 3 & 2 & -2 \end{pmatrix}$$

9.2. By using the Frobenius matrix whose characteristic equation is

$$x^3 - 9{\cdot}40x^2 - 1{\cdot}44x + 13{\cdot}54 = 0$$

find the largest root of this cubic correct to two decimal places.

9.3. (i) A skew–Hermitian matrix is a matrix such that

$$\mathbf{A}^* = -\mathbf{A}$$

Show that the latent roots of a skew–Hermitian matrix are purely imaginary. Hence show that if $\mathbf{A}$ is an $n \times n$ skew–Hermitian matrix where n is odd, then at least one latent root is zero.

(ii) Find the two latent roots of largest magnitude of the matrix

$$\mathbf{A} = \begin{pmatrix} 0 & 3 & 5 & 1 \\ -3 & 0 & 4 & -5 \\ -5 & -4 & 0 & -3 \\ -1 & 5 & 3 & 0 \end{pmatrix}$$

correct to two decimal places.

† See reference 12, Chapter 7.

9.4. Establish the equation

$$\mathbf{Y}_p = k_0 k_1 \dots k_{p-1}[a_1 \lambda_1^p \mathbf{X}_1 + a_2 \lambda_2^p \mathbf{X}_2 + \mathbf{E}(p)]$$

of § 9.2.

9.5. Given that the latent roots of the matrix

$$\mathbf{A} = \begin{pmatrix} 14 & 2 & -12 \\ 4 & 4 & -2 \\ 3 & 2 & -1 \end{pmatrix}$$

are approximately 10, 5 and 1, use the method of § 9.3 to find the largest and smallest of these to two decimal places.

9.6. If

$$\mathbf{A} = \begin{pmatrix} 20 & -10 & 20 \\ -10 & 25 & -5 \\ 20 & -5 & 85 \end{pmatrix} \quad \text{and} \quad \mathbf{LU} = \begin{pmatrix} 10 & 0 & 0 \\ -10 & 5 & 0 \\ 20 & 15 & -10 \end{pmatrix} \begin{pmatrix} 1 & -1 & 2 \\ 0 & 1 & 3 \\ 0 & 0 & 1 \end{pmatrix}$$

show that $\mathbf{A} - 10\mathbf{I} = \mathbf{LU}$. Taking $k_0 = 1$ and

$$\mathbf{Y}_0 = \begin{pmatrix} 1 \\ 0 \\ 0 \end{pmatrix}$$

solve the equations $\mathbf{LUY}_1 = k_0 \mathbf{Y}_0$ by first solving the equations $\mathbf{LZ} = k_0 \mathbf{Y}_0$ where

$$\mathbf{Z} = \mathbf{UY}_1 = \begin{pmatrix} z_1 \\ z_2 \\ z_3 \end{pmatrix}$$

and then solving the equations $\mathbf{UY}_1 = \mathbf{Z}$.

Continue the inverse iteration to find the latent root close to 10 of the matrix $\mathbf{A}$. Notice that $\mathbf{A} - d\mathbf{I}$ only needs to be decomposed into the product $\mathbf{LU}$ once for the method of inverse iteration.

9.7. For the tridiagonal matrix $\mathbf{A}_1$ of example 6.1 find a lower triangular matrix $\mathbf{L}$ with unit elements on its leading diagonal, and an upper triangular matrix $\mathbf{U}$ such that

$$\mathbf{A}_1 - \lambda_1 \mathbf{I} = \mathbf{LU}$$

Solve the equations

$$\mathbf{UY}_1 = \begin{pmatrix} 1 \\ 1 \\ 1 \end{pmatrix}$$

Perform one step of inverse iteration by solving the equations

$$\mathbf{LUY}_2 = k_1 \mathbf{Y}_1$$

to find a more accurate latent vector of $\mathbf{A}_1$.

Finding a starting vector $\mathbf{Y}_1$ by the above is the method recommended by Wilkinson for performing inverse iteration on a tridiagonal matrix. In practice Wilkinson has found that $\mathbf{Y}_3$ is never then needed. For details of the method of triangular decomposition, and the necessary pivoting techniques see (p. 23).

9.8. Given that one latent root of the matrix

$$\mathbf{A} = \begin{pmatrix} -10 & 3 & 3 \\ -50 & 17 & 10 \\ 48 & -19 & -3 \end{pmatrix}$$

is $\lambda_1 = 2$ and that the corresponding latent vector is

$$\mathbf{X}_1 = \begin{pmatrix} 1 \\ 2 \\ 2 \end{pmatrix}$$

use the method of matrix deflation to find the remaining latent roots and vectors.

9.9. The largest latent root of the matrix

$$\mathbf{A} = \begin{pmatrix} 3 & 5 & 3 & 1 \\ 4 & 4 & 3 & 1 \\ 7 & 5 & 6 & -6 \\ 1 & 1 & 1 & 9 \end{pmatrix}$$

is $\lambda_1 = 12$ and the corresponding latent vector is

$$\mathbf{X}_1 = \begin{pmatrix} 1 \\ 1 \\ 1 \\ 1 \end{pmatrix}$$

Use the method of deflation to find a three by three matrix whose latent roots are the remaining latent roots of **A**. By iterating with this three by three matrix find the next largest latent root of **A** and the corresponding latent vector. By deflating once more find the other two latent roots and vectors of **A**.

Chapter 10

THE METHOD OF FRANCIS

The method of Francis, discovered also by Kublanovskaya and generally referred to as the **Q–R** algorithm, is an iterative method that attempts to reduce a matrix to triangular form. From a practical point of view this is one of the most important methods at present available.

10.1 The Iterative Procedure

The method decomposes the matrix **A** into the product

$$\mathbf{A} = \mathbf{Q}_1 \mathbf{U}_1$$

where $\mathbf{Q}_1$ is an orthogonal matrix and $\mathbf{U}_1$ is an upper triangular matrix. Then a matrix $\mathbf{A}_1$ is formed by

$$\mathbf{A}_1 = \mathbf{U}_1 \mathbf{Q}_1$$

and since

$$\mathbf{A} = \mathbf{Q}_1 \mathbf{U}_1 = \mathbf{Q}_1 \mathbf{U}_1 \mathbf{Q}_1 \mathbf{Q}_1^{-1} = \mathbf{Q}_1 \mathbf{A}_1 \mathbf{Q}_1^{-1}$$

$\mathbf{A}_1$ is similar to **A**. The method proceeds iteratively by forming the sequence of matrices $\mathbf{A}_1, \mathbf{A}_2, \ldots, \mathbf{A}_i, \ldots$, where

$$\mathbf{A}_i = \mathbf{U}_i \mathbf{Q}_i$$

$\mathbf{Q}_i$ being orthogonal and $\mathbf{U}_i$ upper triangular and these being given by decomposing $\mathbf{A}_{i-1}$ into the product

$$\mathbf{A}_{i-1} = \mathbf{Q}_i \mathbf{U}_i$$

Under certain conditions $\mathbf{A}_i$ tends towards an upper triangular matrix as $i \to \infty$, so that the latent roots of **A** lie on the diagonal of this matrix.

The complete proof of the **Q–R** algorithm is by no means easy and only an indication as to how the final result may be arrived at is given here.†

10.2 Result of the Iterative Procedure

Let us put

$$\mathbf{W}_i = \mathbf{Q}_1 \mathbf{Q}_2 \ldots \mathbf{Q}_i$$

† For a formal proof see reference 12, § 7.9. Also see reference 7, pp. 515 ff.

Then,

$$\begin{aligned}\mathbf{A}_i &= \mathbf{U}_i\mathbf{Q}_i = \mathbf{Q}_i^{-1}(\mathbf{Q}_i\mathbf{U}_i)\mathbf{Q}_i = \mathbf{Q}_i^{-1}\mathbf{A}_{i-1}\mathbf{Q}_i\\ &= \mathbf{Q}_i^{-1}\mathbf{Q}_{i-1}^{-1}\mathbf{A}_{i-2}\mathbf{Q}_{i-1}\mathbf{Q}_i\\ &= \mathbf{Q}_i^{-1}\mathbf{Q}_{i-1}^{-1}\dots\mathbf{Q}_1^{-1}\mathbf{A}\mathbf{Q}_1\dots\mathbf{Q}_{i-1}\mathbf{Q}_i\\ &= (\mathbf{Q}_1\mathbf{Q}_2\dots\mathbf{Q}_{i-1}\mathbf{Q}_i)^{-1}\mathbf{A}(\mathbf{Q}_1\mathbf{Q}_2\dots\mathbf{Q}_{i-1}\mathbf{Q}_i)\\ &= \mathbf{W}_i^{-1}\mathbf{A}\mathbf{W}_i \qquad (10.1)\end{aligned}$$

so that $\mathbf{A}_i$ is similar to $\mathbf{A}$.

Also,

$$\mathbf{A} = \mathbf{Q}_1\mathbf{U}_1$$

so that

$$\mathbf{A}\mathbf{Q}_1 = \mathbf{Q}_1\mathbf{U}_1\mathbf{Q}_1 = \mathbf{Q}_1\mathbf{A}_1 = \mathbf{Q}_1\mathbf{Q}_2\mathbf{U}_2$$

and

$$\mathbf{A}\mathbf{Q}_1\mathbf{Q}_2 = \mathbf{Q}_1\mathbf{Q}_2\mathbf{U}_2\mathbf{Q}_2 = \mathbf{Q}_1\mathbf{Q}_2\mathbf{A}_2 = \mathbf{Q}_1\mathbf{Q}_2\mathbf{Q}_3\mathbf{U}_3$$

Proceeding in this way we get

$$\mathbf{A}\mathbf{Q}_1\mathbf{Q}_2\dots\mathbf{Q}_i = \mathbf{Q}_1\mathbf{Q}_2\dots\mathbf{Q}_{i+1}\mathbf{U}_{i+1}$$

or

$$\mathbf{A}\mathbf{W}_i = \mathbf{W}_{i+1}\mathbf{U}_{i+1} \qquad (10.2)$$

From equation (10.1) this gives

$$\mathbf{A}_i = \mathbf{W}_i^{-1}\mathbf{A}\mathbf{W}_i = \mathbf{W}_i^{-1}\mathbf{W}_{i+1}\mathbf{U}_{i+1} = \mathbf{W}_i^{T}\mathbf{W}_{i+1}\mathbf{U}_{i+1} \qquad (10.3)$$

since $\mathbf{W}_i$ is orthogonal.

Equation (10.2) allows us to investigate the limit of $\mathbf{W}_i$ as $i \to \infty$, and equation (10.3) then allows us to investigate $\mathbf{A}_i$ as $i \to \infty$. Clearly if $\mathbf{W}_i \to \mathbf{W}$ (say) as $i \to \infty$, then

$$\mathbf{A}_i \to \mathbf{U}_{i+1} \quad \text{as } i \to \infty$$

where $\mathbf{U}_{i+1}$ is of course upper triangular. It is possible to investigate the limit in this full form† but instead an indication is given as to how this may be arrived at inductively.

Instead of considering the sequence

$$\mathbf{A}\mathbf{W}_i = \mathbf{W}_{i+1}\mathbf{U}_{i+1}$$

first consider the equation

$$\mathbf{A}\mathbf{Y}_i = \mathbf{Y}_{i+1}r_{i+1}$$

where $\mathbf{Y}_i$ is a column vector such that $\mathbf{Y}_i^T\mathbf{Y}_i = 1$ and r_i is a scalar.

† See reference 12, § 7.9.

Theorem 10.1

If the latent roots of $\mathbf{A}$ are such that

$$|\lambda_1| > |\lambda_2| \geqslant \ldots \geqslant |\lambda_n|$$

then $r_i \to \lambda_1$ and $\mathbf{Y}_i \to \mathbf{X}_1$ as $i \to \infty$, where $\mathbf{X}_1$ is the latent vector of $\mathbf{A}$ associated with λ_1 and

$$\mathbf{A}\mathbf{Y}_i = r_{i+1}\mathbf{Y}_{i+1}$$

with $\mathbf{Y}_i^T\mathbf{Y}_i = 1$.

Proof

Since r_i is chosen so that $\mathbf{Y}_i^T\mathbf{Y}_i = 1$, the result follows immediately from § 9.1.

Next we consider the iterative procedure

$$\mathbf{A}\mathbf{Y}_i = \mathbf{Y}_{i+1}\mathbf{R}_{i+1}$$

where $\mathbf{Y}_i$ is a matrix containing two columns $\mathbf{Y}_{i1}$ and $\mathbf{Y}_{i2}$ such that $\mathbf{Y}_i^T\mathbf{Y}_i = \mathbf{I}$, and $\mathbf{R}_{i+1}$ is a two by two upper triangular matrix, that is,

$$\mathbf{A}\mathbf{Y}_i = \mathbf{A}(\,\mathbf{Y}_{i1} \quad \mathbf{Y}_{i2}\,) = (\,\mathbf{Y}_{i+1,1} \quad \mathbf{Y}_{i+1,2}\,)\begin{pmatrix} r_{i+1,1} & r_{i+1,2} \\ 0 & r_{i+1,3} \end{pmatrix} = \mathbf{Y}_{i+1}\mathbf{R}_{i+1} \tag{10.4}$$

Theorem 10.2

If the latent roots of $\mathbf{A}$ are such that

$$|\lambda_1| > |\lambda_2| > |\lambda_3| \geqslant \ldots \geqslant |\lambda_n|$$

then $r_{i1} \to \lambda_1$, $r_{i3} \to \lambda_2$ and $\mathbf{Y}_{i1} \to \mathbf{X}_1$ as $i \to \infty$, where these are given by the above iterative procedure of equation (10.4).

Proof

Since $\mathbf{Y}_i^T\mathbf{Y}_i = \mathbf{I}$ we get that $\mathbf{Y}_{i1}^T\mathbf{Y}_{i1} = 1$ and also we have, from equation (10.4), that

$$\mathbf{A}\mathbf{Y}_{i1} = \mathbf{Y}_{i+1,1}r_{i+1,1} + \mathbf{Y}_{i+1,2} \times 0 = r_{i+1,1}\mathbf{Y}_{i+1,1}$$

Hence from theorem 10.1 $r_{i1} \to \lambda_1$ and $\mathbf{Y}_{i1} \to \mathbf{X}_1$ as $i \to \infty$.

The second equation arising from equation (10.4) gives

$$\begin{aligned}\mathbf{A}\mathbf{Y}_{i2} &= \mathbf{Y}_{i+1,1}r_{i+1,2} + \mathbf{Y}_{i+1,2}r_{i+1,3} \\ &= r_{i+1,2}\mathbf{Y}_{i+1,1} + r_{i+1,3}\mathbf{Y}_{i+1,2}\end{aligned}$$

Because $\mathbf{Y}_{i+1}^T\mathbf{Y}_{i+1} = \mathbf{I}$, we must have that $\mathbf{Y}_{i+1,1}^T\mathbf{Y}_{i+1,2} = 0$ and hence $\mathbf{Y}_{i+1,1}$ and $\mathbf{Y}_{i+1,2}$ are linearly independent.

Now $\mathbf{Y}_{i+1,1} \to \mathbf{X}_1$ as $i \to \infty$, so that $\mathbf{Y}_{i+1,2}$ must become linearly independent of $\mathbf{X}_1$ as $i \to \infty$ and therefore becomes dominated by the next largest root, that is λ_2. Hence $r_{i+1,3} \to \lambda_2$ as $i \to \infty$, and the theorem is established.

Notice that $\mathbf{Y}_{i+1,2} \to \mathbf{X}_2$ as $i \to \infty$ only if $\mathbf{X}_1$ and $\mathbf{X}_2$ are orthogonal. In particular this will be true if $\mathbf{A}$ is symmetric. Since $\mathbf{Y}_{i+1,1}$ tends to a limit and $\mathbf{Y}^T_{i+1,1}\mathbf{Y}_{i+1,2} = 0$, $\mathbf{Y}_{i+1,2}$ must also always tend to a limit.

It is hoped it is now clear how the following theorem may be established.

Theorem 10.3

If the latent roots of $\mathbf{A}$ are such that

$$|\lambda_1| > |\lambda_2| > \ldots > |\lambda_n|$$

then in the sequence defined by

$$\mathbf{A}\mathbf{W}_i = \mathbf{W}_{i+1}\mathbf{U}_{i+1}$$

where $\mathbf{W}_i$ is an orthogonal matrix and $\mathbf{U}_i$ is upper triangular,

$$\mathbf{W}_i \to \mathbf{W} \quad \text{(i.e. } \mathbf{W} \text{ tends to a limit)}$$

and the elements on the leading diagonal of $\mathbf{U}_i$ tend to the latent roots of $\mathbf{A}$ as $i \to \infty$.

If $\mathbf{W}_i \to \mathbf{W}$, then from equation (10.3) we get

$$\mathbf{A}_i = \mathbf{W}_i^T\mathbf{W}_{i+1}\mathbf{U}_{i+1} \to \mathbf{W}^T\mathbf{W}\mathbf{U}_{i+1} = \mathbf{U}_{i+1}$$

We have now established that if no two latent roots of $\mathbf{A}$ are of equal modulus then $\mathbf{A}_i$ tends to an upper triangular matrix.

If the latent roots of $\mathbf{A}$ are not all of distinct modulus then $\mathbf{A}_i$ may not tend to an upper triangular matrix, but may instead have a block of elements centred on the leading diagonal whose latent roots correspond to the latent roots of $\mathbf{A}$ of equal modulus.

For example, suppose that $\mathbf{A}$ is a four by four matrix with

$$|\lambda_1| > |\lambda_2| = |\lambda_3| > |\lambda_4|$$

and that

$$\mathbf{W}_i = (\,\mathbf{W}_{i1} \quad \mathbf{W}_{i2} \quad \mathbf{W}_{i3} \quad \mathbf{W}_{i4}\,)$$

Then clearly $\mathbf{W}_{i1} \to \mathbf{X}_1$ and $\mathbf{W}_{i4} \to \mathbf{V}$ (say) as $\mathbf{i} \to \infty$, but $\mathbf{W}_{i2}$ and $\mathbf{W}_{i3}$ may not tend to a limit. (Compare with § 9.2.) Also we can see that

$$\mathbf{U}_i \to \begin{pmatrix} \lambda_1 & x_1 & x_2 & x_3 \\ 0 & a_i & x_4 & x_5 \\ 0 & 0 & b_i & x_6 \\ 0 & 0 & 0 & \lambda_4 \end{pmatrix} \quad \text{as } i \to \infty$$

where $x_1, x_2, \ldots, x_6$ may or may not depend on i.

Then, since

$$\lim_{i\to\infty} \mathbf{W}_i^T \mathbf{W}_{i+1} = \begin{pmatrix} 1 & 0 & 0 & 0 \\ 0 & \mathbf{W}_{i2}^T \mathbf{W}_{i+1,2} & \mathbf{W}_{i2}^T \mathbf{W}_{i+1,3} & 0 \\ 0 & \mathbf{W}_{i3}^T \mathbf{W}_{i+1,2} & \mathbf{W}_{i3}^T \mathbf{W}_{i+1,3} & 0 \\ 0 & 0 & 0 & 1 \end{pmatrix}$$

$\lim_{i\to\infty} \mathbf{A}_i = \lim_{i\to\infty}(\mathbf{W}_i^T \mathbf{W}_{i+1} \mathbf{U}_{i+1})$ is of the form

$$\begin{pmatrix} \lambda_1 & x_1 & x_2 & x_3 \\ 0 & p_i & q_i & y_1 \\ 0 & r_i & s_i & y_2 \\ 0 & 0 & 0 & \lambda_4 \end{pmatrix}$$

and the latent roots λ_2 and λ_3 are the latent roots of the matrix

$$\begin{pmatrix} p_i & q_i \\ r_i & s_i \end{pmatrix}$$

The element r_i may of course be zero, but, in particular, if λ_2 and λ_3 are complex then r_i will not be zero. The extension to the general case should now be clear.

Theorem 10.4

Define a sequence by

$$\mathbf{A}\mathbf{W}_i = \mathbf{W}_{i+1} \mathbf{U}_{i+1}$$

where $\mathbf{W}_i$ is orthogonal and $\mathbf{U}_i$ is upper triangular. Then if

$$\mathbf{A}_i = \mathbf{W}_i^T \mathbf{W}_{i+1} \mathbf{U}_{i+1}$$

$\mathbf{A}_i$ tends towards a block triangular matrix as $i \to \infty$, where the latent roots of each block correspond to the latent roots of equal modulus of the matrix **A**.

10.3 Performing the Method

We now look at how that **Q–R** algorithm is actually performed. We wish to decompose $\mathbf{A}_{i-1}$ into the product

$$\mathbf{A}_{i-1} = \mathbf{Q}_i \mathbf{U}_i$$

This is done by finding $\mathbf{Q}_i$ such that

$$\mathbf{Q}_i^T \mathbf{A}_{i-1} = \mathbf{U}_i$$

We form $\mathbf{Q}_i^T$ as the product of orthogonal matrices which are chosen as in either the Givens or the Householder methods. Taking as representative the three by three matrix,

$$\mathbf{A}_{i-1} = \begin{pmatrix} a_{11} & a_{12} & a_{13} \\ a_{21} & a_{22} & a_{23} \\ a_{31} & a_{32} & a_{33} \end{pmatrix}$$

we wish to introduce zeros into the a_{21}, a_{31} and a_{32} positions. The method of Givens suggests that we rotate in the (1, 2)-, (1, 3)- and (2, 3)-planes, so that the first stage forms

$$\mathbf{P}_1\mathbf{A}_{i-1} = \begin{pmatrix} c_1 & s_1 & 0 \\ -s_1 & c_1 & 0 \\ 0 & 0 & 1 \end{pmatrix} \begin{pmatrix} a_{11} & a_{12} & a_{13} \\ a_{21} & a_{22} & a_{23} \\ a_{31} & a_{32} & a_{33} \end{pmatrix}$$

where $c_1 = a_{11}(a_{11}^2 + a_{21}^2)^{-\frac{1}{2}}$ and $s_1 = a_{21}(a_{11}^2 + a_{21}^2)^{-\frac{1}{2}}$ which will introduce the required zero into the a_{21} position. Then successively $\mathbf{P}_2(\mathbf{P}_1\mathbf{A}_{i-1})$ and $\mathbf{P}_3(\mathbf{P}_2\mathbf{P}_1\mathbf{A}_{i-1})$ are formed, where $\mathbf{P}_2$ and $\mathbf{P}_3$ are of the form

$$\mathbf{P}_2 = \begin{pmatrix} c_2 & 0 & s_2 \\ 0 & 1 & 0 \\ -s_2 & 0 & c_2 \end{pmatrix}, \quad \mathbf{P}_3 = \begin{pmatrix} 1 & 0 & 0 \\ 0 & c_3 & s_3 \\ 0 & -s_3 & c_3 \end{pmatrix}$$

so that $\mathbf{Q}_i^T = \mathbf{P}_3\mathbf{P}_2\mathbf{P}_1$. Then $\mathbf{A}_i$ is given by

$$\mathbf{A}_i = \mathbf{U}_i\mathbf{Q}_i = \mathbf{U}_i(\mathbf{P}_3\mathbf{P}_2\mathbf{P}_1)^T = \mathbf{U}_i\mathbf{P}_1^T\mathbf{P}_2^T\mathbf{P}_3^T$$

In forming $\mathbf{A}_i$ we compute in sequence $\mathbf{U}_i\mathbf{P}_1^T$, $(\mathbf{U}_i\mathbf{P}_1^T)\mathbf{P}_2^T$, $(\mathbf{U}_i\mathbf{P}_1^T\mathbf{P}_2^T)\mathbf{P}_3^T$. We can of course replace the Givens type transformations with those of Householder and for computer purposes this is undoubtedly preferable in general.

Example 10.1

$$\mathbf{A} = \begin{pmatrix} 3 & 7 \\ 4 & 6 \end{pmatrix}$$

Then

$$c = a_{11}(a_{11}^2 + a_{21}^2)^{-\frac{1}{2}} = \tfrac{3}{5}, \quad s = a_{21}(a_{11}^2 + a_{21}^2)^{-\frac{1}{2}} = \tfrac{4}{5}$$

so that

$$\mathbf{Q}_1^T\mathbf{A} = \begin{pmatrix} \frac{3}{5} & \frac{4}{5} \\ -\frac{4}{5} & \frac{3}{5} \end{pmatrix} \begin{pmatrix} 3 & 7 \\ 4 & 6 \end{pmatrix} = \begin{pmatrix} 5 & 9 \\ 0 & -2 \end{pmatrix} = \mathbf{U}_1$$

Then,

$$\mathbf{A}_1 = \mathbf{U}_1\mathbf{Q}_1 = \begin{pmatrix} 5 & 9 \\ 0 & -2 \end{pmatrix} \begin{pmatrix} \frac{3}{5} & -\frac{4}{5} \\ \frac{4}{5} & \frac{3}{5} \end{pmatrix} = \begin{pmatrix} 10{\cdot}2 & 1{\cdot}4 \\ -1{\cdot}6 & -1{\cdot}2 \end{pmatrix}$$

At this stage $\lambda_1 \simeq 10{\cdot}2$ and $\lambda_2 \simeq -1{\cdot}2$.

Now

$$c = 10{\cdot}2(10{\cdot}2^2 + (-1{\cdot}6)^2)^{-\frac{1}{2}} = 0{\cdot}99$$

$$s = -1{\cdot}6(10{\cdot}2^2 + (-1{\cdot}6)^2)^{-\frac{1}{2}} = -0{\cdot}15$$

so that

$$\mathbf{Q}_2^T\mathbf{A}_1 = \begin{pmatrix} 0{\cdot}99 & -0{\cdot}15 \\ 0{\cdot}15 & 0{\cdot}99 \end{pmatrix} \begin{pmatrix} 10{\cdot}2 & 1{\cdot}4 \\ -1{\cdot}6 & -1{\cdot}2 \end{pmatrix} = \begin{pmatrix} 10{\cdot}34 & 1{\cdot}57 \\ 0 & -0{\cdot}98 \end{pmatrix} = \mathbf{U}_2$$

Then

$$\mathbf{U}_2\mathbf{Q}_2 = \begin{pmatrix} 10{\cdot}34 & 1{\cdot}57 \\ 0 & -0{\cdot}98 \end{pmatrix} \begin{pmatrix} 0{\cdot}99 & 0{\cdot}15 \\ -0{\cdot}15 & 0{\cdot}99 \end{pmatrix} = \begin{pmatrix} 10{\cdot}00 & 3{\cdot}11 \\ 0{\cdot}15 & -0{\cdot}97 \end{pmatrix} = \mathbf{A}_2$$

$$\lambda_1 \simeq 10{\cdot}00, \quad \lambda_2 \simeq -0{\cdot}97$$

$$c = 10{\cdot}00(10{\cdot}00^2 + 0{\cdot}15^2)^{-\frac{1}{2}} = 1{\cdot}00, \quad s = 0{\cdot}15(1{\cdot}00^2 + 0{\cdot}15^2)^{-\frac{1}{2}} = 0{\cdot}01$$

so that

$$\mathbf{Q}_3^T\mathbf{A}_2 = \begin{pmatrix} 1{\cdot}00 & 0{\cdot}01 \\ -0{\cdot}01 & 1{\cdot}00 \end{pmatrix} \begin{pmatrix} 10{\cdot}00 & 3{\cdot}11 \\ 0{\cdot}15 & -0{\cdot}97 \end{pmatrix} = \begin{pmatrix} 10{\cdot}00 & 3{\cdot}10 \\ 0 & -1{\cdot}00 \end{pmatrix} = \mathbf{U}_3$$

Then

$$\mathbf{U}_3\mathbf{Q}_3 = \begin{pmatrix} 10{\cdot}00 & 3{\cdot}10 \\ 0 & -1{\cdot}00 \end{pmatrix} \begin{pmatrix} 1{\cdot}00 & -0{\cdot}01 \\ 0{\cdot}01 & 1{\cdot}00 \end{pmatrix} = \begin{pmatrix} 10{\cdot}03 & 3{\cdot}00 \\ -0{\cdot}01 & -1{\cdot}00 \end{pmatrix} = \mathbf{A}_3$$

$$\lambda_1 \simeq 10{\cdot}03, \quad \lambda_2 \simeq -1{\cdot}00$$

These are good approximations to the exact roots which are $\lambda_1 = 10$ and $\lambda_2 = -1$. Notice that rounding errors have not seriously affected convergence to the roots. Notice also that

$$\mathbf{W}_2 = \mathbf{Q}_1\mathbf{Q}_2 = \begin{pmatrix} 0{\cdot}6 & -0{\cdot}8 \\ 0{\cdot}8 & 0{\cdot}6 \end{pmatrix} \begin{pmatrix} 0{\cdot}99 & 0{\cdot}15 \\ -0{\cdot}15 & 0{\cdot}99 \end{pmatrix} = \begin{pmatrix} 0{\cdot}71 & -0{\cdot}70 \\ 0{\cdot}70 & 0{\cdot}71 \end{pmatrix}$$

and

$$\mathbf{W}_3 = \mathbf{Q}_1\mathbf{Q}_2\mathbf{Q}_3 = \begin{pmatrix} 0{\cdot}71 & -0{\cdot}70 \\ 0{\cdot}70 & 0{\cdot}71 \end{pmatrix} \begin{pmatrix} 1{\cdot}00 & -0{\cdot}01 \\ 0{\cdot}01 & 1{\cdot}00 \end{pmatrix} = \begin{pmatrix} 0{\cdot}70 & -0{\cdot}71 \\ 0{\cdot}71 & 0{\cdot}70 \end{pmatrix}$$

Also

$$\mathbf{X}_1 = \begin{pmatrix} \frac{1}{\sqrt{2}} \\ \frac{1}{\sqrt{2}} \end{pmatrix} \simeq \begin{pmatrix} 0{\cdot}707 \\ 0{\cdot}707 \end{pmatrix}$$

Actually in exact arithmetic we clearly have that

$$\mathbf{W}_i \to \begin{pmatrix} \frac{1}{\sqrt{2}} & -\frac{1}{\sqrt{2}} \\ \frac{1}{\sqrt{2}} & \frac{1}{\sqrt{2}} \end{pmatrix} \quad \text{and} \quad \mathbf{U}_i \to \begin{pmatrix} 10 & 3 \\ 0 & -1 \end{pmatrix} \quad \text{as} \quad i \to \infty$$

Example 10.2

$$\mathbf{A} = \begin{pmatrix} 4 & -3 \\ 3 & -2 \end{pmatrix}$$

Then

$$c = \tfrac{4}{5}, \quad s = \tfrac{3}{5}$$

so that

$$\mathbf{Q}_1^T \mathbf{A} = \begin{pmatrix} 0{\cdot}8 & 0{\cdot}6 \\ -0{\cdot}6 & 0{\cdot}8 \end{pmatrix} \begin{pmatrix} 4 & -3 \\ 3 & -2 \end{pmatrix} = \begin{pmatrix} 5{\cdot}0 & -3{\cdot}6 \\ 0 & 0{\cdot}2 \end{pmatrix} = \mathbf{U}_1$$

and

$$\mathbf{U}_1 \mathbf{Q}_1 = \begin{pmatrix} 5{\cdot}0 & -3{\cdot}6 \\ 0 & 0{\cdot}2 \end{pmatrix} \begin{pmatrix} 0{\cdot}8 & -0{\cdot}6 \\ 0{\cdot}6 & 0{\cdot}8 \end{pmatrix} = \begin{pmatrix} 1{\cdot}84 & -5{\cdot}88 \\ 0{\cdot}12 & 0{\cdot}16 \end{pmatrix} = \mathbf{A}_1$$

$$\lambda_1 \simeq 1{\cdot}84, \quad \lambda_2 \simeq 0{\cdot}16$$

$$c = 0{\cdot}99, \quad s = 0{\cdot}07$$

so that

$$\mathbf{Q}_2^T \mathbf{A}_1 = \begin{pmatrix} 0{\cdot}99 & 0{\cdot}07 \\ -0{\cdot}07 & 0{\cdot}99 \end{pmatrix} \begin{pmatrix} 1{\cdot}84 & -5{\cdot}88 \\ 0{\cdot}12 & 0{\cdot}16 \end{pmatrix} = \begin{pmatrix} 1{\cdot}83 & -5{\cdot}81 \\ 0 & 0{\cdot}57 \end{pmatrix} = \mathbf{U}_2$$

and

$$\mathbf{U}_2 \mathbf{Q}_2 = \begin{pmatrix} 1{\cdot}83 & -5{\cdot}81 \\ 0 & 0{\cdot}57 \end{pmatrix} \begin{pmatrix} 0{\cdot}99 & -0{\cdot}07 \\ 0{\cdot}07 & 0{\cdot}99 \end{pmatrix} = \begin{pmatrix} 1{\cdot}41 & -5{\cdot}88 \\ 0{\cdot}04 & 0{\cdot}56 \end{pmatrix} = \mathbf{A}_2$$

$$\lambda_1 \simeq 1{\cdot}41, \quad \lambda_2 \simeq 0{\cdot}56$$

$$c = 1{\cdot}00, \quad s = 0{\cdot}03$$

so that

$$\mathbf{Q}_3^T \mathbf{A}_2 = \begin{pmatrix} 1{\cdot}00 & 0{\cdot}03 \\ -0{\cdot}03 & 1{\cdot}00 \end{pmatrix} \begin{pmatrix} 1{\cdot}41 & -5{\cdot}88 \\ 0{\cdot}04 & 0{\cdot}56 \end{pmatrix} = \begin{pmatrix} 1{\cdot}41 & -5{\cdot}86 \\ 0 & 0{\cdot}74 \end{pmatrix} = \mathbf{U}_3$$

and

$$\mathbf{U}_3 \mathbf{Q}_3 = \begin{pmatrix} 1{\cdot}41 & -5{\cdot}86 \\ 0 & 0{\cdot}74 \end{pmatrix} \begin{pmatrix} 1{\cdot}00 & -0{\cdot}03 \\ 0{\cdot}03 & 1{\cdot}00 \end{pmatrix} = \begin{pmatrix} 1{\cdot}23 & -5{\cdot}90 \\ 0{\cdot}02 & 0{\cdot}74 \end{pmatrix} = \mathbf{A}_3$$

$$\lambda_1 \simeq 1{\cdot}23, \quad \lambda_2 \simeq 0{\cdot}74$$

Hence convergence is slow to the latent roots which are $\lambda_1 = \lambda_2 = 1$. But if **A** had been, say, a four by four matrix with the other two latent roots, for example, $\lambda_3 = 10$ and $\lambda_4 = 5$, then at this stage we might have had

$$\mathbf{A}_3 = \begin{pmatrix} 10+\varepsilon_1 & x_1 & x_2 & x_3 \\ 0 & 5+\varepsilon_2 & x_4 & x_5 \\ 0 & 0 & 1{\cdot}23 & -5{\cdot}90 \\ 0 & 0 & 0{\cdot}02 & 0{\cdot}74 \end{pmatrix}$$

where ε_1 and ε_2 are small.

10.4 The Q–R Algorithm and Hessenberg Form

Clearly for a general matrix larger than a two by two matrix the **Q–R** algorithm involves an excessive number of calculations. For this reason it is advisable to reduce first the matrix to an upper Hessenberg form, that is, a matrix of the form

$$\mathbf{B} = \begin{pmatrix} b_{11} & b_{12} & b_{13} & b_{14} & \dots & b_{1n} \\ b_{21} & b_{22} & b_{23} & b_{24} & \dots & b_{2n} \\ 0 & b_{32} & b_{33} & b_{34} & \dots & b_{3n} \\ 0 & 0 & b_{43} & b_{44} & \dots & b_{4n} \\ \vdots & \vdots & \vdots & \vdots & & \vdots \\ 0 & 0 & 0 & 0 & \dots & b_{nn} \end{pmatrix}$$

We can apply either the Givens or the Householder transformations to a matrix to obtain the Hessenberg form. Of course, if the matrix is symmetric then **B** is tridiagonal. The important point here is that the **Q–R** algorithm preserves the Hessenberg form, which makes this an extremely useful technique.

Example 10.3

$$\mathbf{B} = \begin{pmatrix} 4 & 27 & 1 \\ 3 & 24 & -3 \\ 0 & 4 & 5 \end{pmatrix}$$

Then,

$$c_1 = 4(4^2+3^2)^{-\frac{1}{2}} = \tfrac{4}{5}, \quad s_1 = 3(4^2+3^2)^{-\frac{1}{2}} = \tfrac{3}{5}$$

so that

$$\mathbf{P}_1\mathbf{B} = \begin{pmatrix} \frac{4}{5} & \frac{3}{5} & 0 \\ -\frac{3}{5} & \frac{4}{5} & 0 \\ 0 & 0 & 1 \end{pmatrix} \begin{pmatrix} 4 & 27 & 1 \\ 3 & 24 & -3 \\ 0 & 4 & 5 \end{pmatrix} = \begin{pmatrix} 5 & 36 & -1 \\ 0 & 3 & -3 \\ 0 & 4 & 5 \end{pmatrix}$$

$$c_3 = 3(3^2+4^2)^{-\frac{1}{2}} = \tfrac{3}{5}, \quad s_3 = 4(3^2+4^2)^{-\frac{1}{2}} = \tfrac{4}{5}$$

so that

$$\mathbf{P}_3\mathbf{P}_1\mathbf{B} = \begin{pmatrix} 1 & 0 & 0 \\ 0 & \frac{3}{5} & \frac{4}{5} \\ 0 & -\frac{4}{5} & \frac{3}{5} \end{pmatrix} \begin{pmatrix} 5 & 36 & -1 \\ 0 & 3 & -3 \\ 0 & 4 & 5 \end{pmatrix}$$

$$= \begin{pmatrix} 5{\cdot}0 & 36{\cdot}0 & -1{\cdot}0 \\ 0 & 5{\cdot}0 & 2{\cdot}2 \\ 0 & 0 & 5{\cdot}6 \end{pmatrix} = \mathbf{Q}_1^T\mathbf{B} = \mathbf{U}_1$$

Then

$$\mathbf{U}_1\mathbf{P}_1^T = \begin{pmatrix} 5{\cdot}0 & 36{\cdot}0 & -1{\cdot}0 \\ 0 & 5{\cdot}0 & 2{\cdot}2 \\ 0 & 0 & 5{\cdot}6 \end{pmatrix} \begin{pmatrix} \frac{4}{5} & -\frac{3}{5} & 0 \\ \frac{3}{5} & \frac{4}{5} & 0 \\ 0 & 0 & 1 \end{pmatrix} = \begin{pmatrix} 25{\cdot}6 & 25{\cdot}8 & -1{\cdot}0 \\ 3{\cdot}0 & 4{\cdot}0 & 2{\cdot}2 \\ 0 & 0 & 5{\cdot}6 \end{pmatrix}$$

and

$$\mathbf{U}_1\mathbf{P}_1^T\mathbf{P}_3^T = \begin{pmatrix} 25{\cdot}6 & 25{\cdot}8 & -1{\cdot}0 \\ 3{\cdot}0 & 4{\cdot}0 & 2{\cdot}2 \\ 0 & 0 & 5{\cdot}6 \end{pmatrix} \begin{pmatrix} 1 & 0 & 0 \\ 0 & \frac{3}{5} & -\frac{4}{5} \\ 0 & \frac{4}{5} & \frac{3}{5} \end{pmatrix}$$

$$= \begin{pmatrix} 25{\cdot}60 & 14{\cdot}68 & -21{\cdot}24 \\ 3{\cdot}00 & 4{\cdot}16 & -1{\cdot}88 \\ 0 & 4{\cdot}48 & 3{\cdot}36 \end{pmatrix} = \mathbf{U}_1\mathbf{Q}_1 = \mathbf{B}_1$$

$$\lambda_1 \simeq 25{\cdot}6, \quad \lambda_2 \simeq 4{\cdot}16, \quad \lambda_3 \simeq 3{\cdot}36$$

and we see that $\mathbf{B}_1$ is still a Hessenberg matrix. Clearly from the way in which it is formed this will always be so.

Working to two decimal places, the next two iterations yield

$$\mathbf{B}_2 = \begin{pmatrix} 27{\cdot}25 & -13{\cdot}15 & -20{\cdot}37 \\ 0{\cdot}61 & 5{\cdot}24 & -2{\cdot}86 \\ 0 & 0{\cdot}85 & 0{\cdot}46 \end{pmatrix}, \quad \mathbf{B}_3 = \begin{pmatrix} 27{\cdot}00 & -16{\cdot}61 & -18{\cdot}69 \\ 0{\cdot}11 & 5{\cdot}16 & -3{\cdot}17 \\ 0 & 0{\cdot}12 & 0{\cdot}81 \end{pmatrix}$$

The exact latent roots are $\lambda_1 = 27$, $\lambda_2 = 5$, $\lambda_3 = 1$. Notice that convergence to the largest latent root has already taken place. It is possible to improve convergence to the next root by using a shift of origin similar to that of § 9.3.

10.5 Shift of Origin

Consider the sequence

$$\mathbf{A}_{i+1} = \mathbf{U}_i\,\mathbf{Q}_i + p_i\,\mathbf{I}$$

where

$$\mathbf{Q}_i\,\mathbf{U}_i = \mathbf{A}_i - p_i\,\mathbf{I}$$

Then

$$\begin{aligned}\mathbf{A}_{i+1} &= \mathbf{U}_i\,\mathbf{Q}_i + p_i\,\mathbf{I} = \mathbf{Q}_i^{-1}\,\mathbf{Q}_i\,\mathbf{U}_i\,\mathbf{Q}_i + p_i\,\mathbf{I} = \mathbf{Q}_i^{-1}(\mathbf{A}_i - p_i\,\mathbf{I})\,\mathbf{Q}_i + p_i\,\mathbf{I} \\ &= \mathbf{Q}_i^{-1}\,\mathbf{A}_i\,\mathbf{Q}_i - p_i\,\mathbf{Q}_i^{-1}\,\mathbf{I}\mathbf{Q}_i + p_i\,\mathbf{I} = \mathbf{Q}_i^{-1}\,\mathbf{A}_i\,\mathbf{Q}_i - p_i\,\mathbf{I} + p_i\,\mathbf{I} \\ &= \mathbf{Q}_i^{-1}\,\mathbf{A}_i\,\mathbf{Q}_i\end{aligned}$$

so that $\mathbf{A}_{i+1}$ is still similar to $\mathbf{A}_i$ and must then, of course, be similar to $\mathbf{A}$. The standard **Q–R** algorithm takes $p_i = 0$ for all i. A suitable choice of p_i may improve convergence to particular latent roots. For example, in the matrix $\mathbf{B}_3$ of example 10.1 a possible choice to improve convergence to $\lambda_2 = 5$ would be to take $p_3 = 0{\cdot}81$, this being the current estimate of λ_3. Whereas the latent roots of $\mathbf{B}_3$ are $\lambda_1 \simeq 27$, $\lambda_2 \simeq 5$, $\lambda_3 \simeq 1$, the latent roots of $\mathbf{B}_3 - p_3\,\mathbf{I}$ would obviously be $\lambda_1 \simeq 26{\cdot}19$, $\lambda_2 \simeq 4{\cdot}19$, $\lambda_3 \simeq 0{\cdot}19$, so that we have considerably improved the dominance of λ_2 over λ_3.

Wilkinson gives a full discussion on suitable choices of p_i, and of a powerful double-shift technique.†

10.6 Further Comments on the Q–R Algorithm

As a hand method the **Q–R** algorithm clearly presents a large amount of computation and for this reason only three simple examples were given earlier. As a computer method the **Q–R** algorithm, with suitable shift of origin, is extremely powerful, mainly because it is a very stable method.

If the **Q–R** algorithm is applied to a general matrix then Householder-type transformations are most suitable in reducing **A** to triangular form. But if the matrix **A** is first reduced to Hessenberg form, which is generally advisable, then Givens-type transformations only are needed in reducing **B** to triangular form since each column of **B** only requires the introduction of one zero.

† See reference 7, Chapter 8, §§ 36–45.

10.7 Exercises

10.1. Apply the **Q–R** algorithm to the following matrices:

(i) $\mathbf{A} = \begin{pmatrix} 3{\cdot}0 & 8{\cdot}5 \\ 4{\cdot}0 & 18{\cdot}0 \end{pmatrix}$, (ii) $\mathbf{A} = \begin{pmatrix} 7 & -1 \\ 24 & -7 \end{pmatrix}$

(iii) $\mathbf{B} = \begin{pmatrix} 4 & 57 & 2 \\ 3 & 34 & -3 \\ 0 & 24 & 6 \end{pmatrix}$, (iv) $\mathbf{B} = \begin{pmatrix} 20 & 19 & -29 \\ 15 & 19 & -10 \\ 0 & 3 & -4 \end{pmatrix}$

10.2. Prove theorem 10.3.

10.3. Prove that the **Q–R** algorithm preserves the Hessenberg form.

10.4. If

$$\mathbf{B} = \begin{pmatrix} 0 & 0 & 1 \\ 1 & 0 & 0 \\ 0 & 1 & 0 \end{pmatrix}$$

show that the **Q–R** algorithm gives $\mathbf{B}_i = \mathbf{B}$ for all i. Use a shift of origin to try to obtain convergence.

10.4. Reduce the matrix

$$\mathbf{A} = \begin{pmatrix} 99 & 700 & -70 \\ -14 & -99 & 10 \\ 1 & 7 & 6 \end{pmatrix}$$

to an upper Hessenberg matrix **B**, performing all calculations correct to four significant figures. Apply the **Q–R** algorithm to the matrix **B**, again using four significant figure accuracy, and employing suitable shifts of origin. Compare the latent roots so obtained to those of exercises 3.2 and 8.4.

Chapter 11

OTHER METHODS AND FINAL COMMENTS

11.1 Brief Summary of Other Methods

Some of the other methods available are listed below with very brief comments, and references.

1. *The Method of Rutishauser* which is also called the **L–R** algorithm. It led to the development of the **Q–R** algorithm of Francis and Kublanovskaya.

The matrix $\mathbf{A}$ is decomposed into a lower triangular matrix $\mathbf{L}_1$ and an upper triangular matrix $\mathbf{U}_1$ such that $|\mathbf{L}_1| = 1$ and

$$\mathbf{A} = \mathbf{L}_1 \mathbf{U}_1$$

We then form $\mathbf{A}_2 = \mathbf{U}_1 \mathbf{L}_1$ and $\mathbf{A}_2$ is then similarly decomposed as

$$\mathbf{A}_2 = \mathbf{L}_2 \mathbf{U}_2$$

and $\mathbf{A}_3 = \mathbf{U}_2 \mathbf{L}_2$. This process is continued iteratively and in general the sequence $\mathbf{A}_1, \mathbf{A}_2, \ldots$ will converge to an upper triangular matrix.

The method is important since it led to the **Q–R** algorithm, but it is not as general or as stable as that algorithm. (See reference 7, chapter 8; reference 13, pp. 45 ff.; reference 12, § 7.7; reference 21, pp. 475 ff.)

2. *The Method of Jacobi* is an iterative method, which uses plane rotations similar to those of Givens', but with the aim of producing a diagonal rather than a tridiagonal matrix. It is a stable method and generally produces good latent vectors, but the number of calculations required will generally be large compared with the methods of Givens or Householder. (See reference 7, pp. 266–282.)

3. *The Method of Kaiser* uses a Householder type of transformation, but like the Jacobi method attempts to diagonalize the matrix. The method achieves this by maximizing the first element on the leading diagonal rather than reducing off-diagonal elements to zero. The method seems promising, especially for large matrices when only a few of the latent roots are required. (See reference 14.)

4. *The Leverrier–Fadeev Method* is based on the fact that

$$\sum_{i=1}^{n} \lambda_i^k = \text{Trace of } \mathbf{A}^k$$

It computes, for any matrix **A**, the coefficients of the characteristic equation and the adjoint matrix of **A**. The sequence

$$\mathbf{A}_r = \mathbf{A}(\mathbf{A}_{r-1} - p_{r-1}\mathbf{I})$$

is constructed, where $\mathbf{A}_1 = \mathbf{A}$ and

$$p_r = \frac{\text{Trace of } \mathbf{A}_r}{r}$$

Then the characteristic equation is given by

$$\lambda^n - p_1\lambda^{n-1} - p_2\lambda^{n-2} - \ldots - p_{n-1}\lambda - p_n = 0$$

and the adjoint matrix is $(-1)^{n-1}\mathbf{B}_{n-1}$, where $\mathbf{B}_{n-1} = \mathbf{A}_{n-1} - p_{n-1}\mathbf{I}$. If $\mathbf{A}^{-1}$ exists, then $\mathbf{A}^{-1} = \mathbf{B}_{n-1}/p_n$.

The method has no cases of breakdown, but the number of calculations is somewhat prohibitive. (See reference 15, p. 193; reference 6, p. 177.)

5. *The Escalator Method* finds the relation between the latent roots of the matrix and those of its principal submatrix. Then commencing with a two by two matrix we successively build up to the n by n matrix, finding at each stage the latent roots of the submatrix. The advantage of the method is that accuracy may be fairly easily checked at each stage, but again the number of calculations is prohibitive. (See reference 6, p. 183.)

6. *The Method of Eberlein* is based on the fact that any matrix is similar to a matrix which is arbitrarily close to a normal matrix. (**A** is normal if $\mathbf{A}^*\mathbf{A} = \mathbf{A}\mathbf{A}^*$.) Eberlein's method attempts to reduce **A** to a normal matrix **N**, such that

$$|\mathbf{N}| = |\mathbf{D}_1|\,|\mathbf{D}_2| \ldots |\mathbf{D}_r|$$

where no $\mathbf{D}_i$ is greater than a two by two matrix. The advantage of this method is that the latent root problem of a normal matrix is well-conditioned. Developments along these lines seem likely to provide an excellent algorithm. (See reference 13, p. 53; reference 22.)

7. *Matrix Squaring* is a method similar to the iterative method of § 9.1, but instead of working throughout with **A** or $\mathbf{A}^2$ we work with the sequence of matrices $\mathbf{A}, \mathbf{A}^2, \mathbf{A}^4, \mathbf{A}^8, \ldots, \mathbf{A}^{2r}$. This is useful when the two latent roots of largest modulus are poorly separated. (See reference 7, p. 615.)

8. *Spectroscopic Eigenvalue Analysis* is an iterative method due to Lanczos and is based on replacing equation (9.4)

$$\begin{aligned}\mathbf{Y}_i &= k_{i-1}\mathbf{A}\mathbf{Y}_{i-1}\\ &= k_0 k_1 \ldots k_{i-1}(a_1\lambda_1^i\mathbf{X}_1 + a_2\lambda_2^i\mathbf{X}_2 + \ldots + a_n\lambda_n^i\mathbf{X}_n)\end{aligned}$$

by

$$\mathbf{Y}_i = a_1\mathbf{T}_i(\lambda_1)\mathbf{X}_1 + a_2\mathbf{T}_i(\lambda_2)\mathbf{X}_2 + \ldots + a_n\mathbf{T}_i(\lambda_n)\mathbf{X}_n$$

where $\mathbf{T}_i(\lambda)$ is the ith Chebyshev polynomial of the first kind. This, of course, means scaling $\mathbf{A}$ so that $-1 \leqslant \lambda_i \leqslant 1$. There seems, at present, to be no special advantage in this method over the more usual iterative methods. (See reference 16, p. 190; reference 12, p. 188; reference 7, p. 617.)

11.2 Final Comments

Whenever we use a method that finds directly the characteristic equation, the condition of this equation should be carefully investigated, for many polynomials are extremely ill-conditioned. We may easily turn a well-conditioned latent root problem into an ill-conditioned polynomial problem.† Though there are a large number of methods available, the complete solution of the latent root and vector problem for a large unsymmetric matrix still presents severe difficulties.

To date, the best answer seems to be to use a Householder-type transformation to reduce the matrix to Hessenberg form, and then to apply the **Q–R** algorithm to the Hessenberg matrix, both of these being stable processes.

For a symmetric matrix the method of Householder, together with inverse iteration for finding the latent vectors of the tridiagonal matrix, is generally quite adequate.

If we require only one, or a few, of the latent roots, then it is best to use one of the iterative methods together with matrix deflation when more than one root is wanted. If approximate values of the required latent roots are known then inverse iteration provides an excellent method.

† See reference 13, pp. 28–31.

Appendix 1

THE LATENT ROOTS OF A COMMON TRIDIAGONAL MATRIX

Let **A** be the common tridiagonal $n \times n$ matrix given by

$$\mathbf{A} = \begin{pmatrix} a & c & 0 & 0 & \dots & 0 & 0 \\ b & a & c & 0 & \dots & 0 & 0 \\ 0 & b & a & c & \dots & 0 & 0 \\ \vdots & \vdots & \vdots & \vdots & & \vdots & \vdots \\ 0 & 0 & 0 & 0 & \dots & b & a \end{pmatrix}$$

We can write **A** as

$$\mathbf{A} = a\mathbf{I} + \mathbf{B}$$

so that if λ is a latent root of **A** and β is a latent root of **B**, we have from theorem 1.13 that

$$\lambda = a + \beta \tag{1}$$

If $\mathbf{D}_n(\beta) = |\mathbf{B} - \beta\mathbf{I}|$, we have

$$\mathbf{D}_n(\beta) = \begin{vmatrix} -\beta & c & 0 & 0 & \dots & 0 & 0 \\ b & -\beta & c & 0 & \dots & 0 & 0 \\ 0 & b & -\beta & c & \dots & 0 & 0 \\ \vdots & \vdots & \vdots & \vdots & & \vdots & \vdots \\ 0 & 0 & 0 & 0 & \dots & b & -\beta \end{vmatrix} = 0$$

and expanding $\mathbf{D}_n(\beta)$ by the first row we get

$$\mathbf{D}_n(\beta) = -\beta\mathbf{D}_{n-1}(\beta) - bc\mathbf{D}_{n-2}(\beta) \tag{2}$$

which is very close to the recurrence relation for a Chebyshev polynomial. Bearing this in mind we put

$$\beta = 2(bc)^{\frac{1}{2}}x \quad \text{and} \quad \mathbf{D}_n(\beta) = (-1)^n (bc)^{n/2} \mathbf{U}_n(x)$$

and substituting in equation (2) we get

$$(-1)^n (bc)^{n/2} \mathbf{U}_n(x) = -2(-1)^{n-1} (bc)^{n/2} x\mathbf{U}_{n-1}(x) - (-1)^{n-2} (bc)^{n/2} \mathbf{U}_{n-2}(x)$$

which gives

$$\mathbf{U}_n(x) = 2x\mathbf{U}_{n-1}(x) - \mathbf{U}_{n-2}(x)$$

and this is the recurrence relation for a Chebyshev polynomial. Noting that $\mathbf{D}_0(\beta) = 1$ and $\mathbf{D}_1(\beta) = -\beta$ we see that $\mathbf{U}_n(x)$ is a Chebyshev polynomial of the second kind, and hence

$$\mathbf{U}_n(x) = \frac{\sin(n+1)\theta}{\sin\theta}$$

where $x = \cos\theta$ and we can see that the zeros of $\mathbf{U}_n(x)$ occur when

$$x = \cos\frac{r\pi}{n+1}, \quad r = 1, 2, \ldots, n$$

which means that $\mathbf{D}_n(\beta) = 0$ when

$$\beta = 2\sqrt{(bc)}\cos\frac{r\pi}{n+1}, \quad r = 1, 2, \ldots, n \tag{3}$$

and hence the latent roots of $\mathbf{A}$ are given by

$$\lambda_r = a + 2\sqrt{(bc)}\cos\frac{r\pi}{n+1}, \quad r = 1, 2, \ldots, n$$

No knowledge of Chebyshev polynomials is required. Substitution of equation (3) in (2) shows that these are required roots.

SOLUTIONS TO EXERCISES

Chapter 1

1.1. (i)
$$\lambda_1 = 3, \quad \lambda_2 = 2, \quad \mathbf{X}_1 = k\begin{pmatrix} 5 \\ -4 \end{pmatrix}, \quad \mathbf{X}_2 = k\begin{pmatrix} 1 \\ -1 \end{pmatrix}$$

(ii)
$$\lambda_1, \lambda_2 = a \pm bi, \quad \mathbf{X}_1, \mathbf{X}_2 = k\begin{pmatrix} 1 \\ \mp i \end{pmatrix} \quad (i = \sqrt{-1})$$

(iii)
$$\lambda_1 = 3, \quad \lambda_2 = 2, \quad \lambda_3 = 1$$

$$\mathbf{X}_1 = k\begin{pmatrix} 1 \\ 1 \\ 1 \end{pmatrix}, \quad \mathbf{X}_2 = k\begin{pmatrix} 1 \\ 2 \\ 1 \end{pmatrix}, \quad \mathbf{X}_3 = k\begin{pmatrix} 1 \\ 13 \\ 3 \end{pmatrix}$$

(iv)
$$\lambda_1 = \lambda_2 = \lambda_3 = 1, \quad \mathbf{X} = k\begin{pmatrix} 1 \\ 0 \\ 1 \end{pmatrix}$$

1.5. Sinh θ and cosh θ can be defined from Fig. 5 as

$$\sinh\theta = \frac{y}{r}, \qquad \cosh\theta = \frac{x}{r}$$

where the hyperbolic angle θ is defined as $\theta = 2A/r^2$, A being the shaded area.

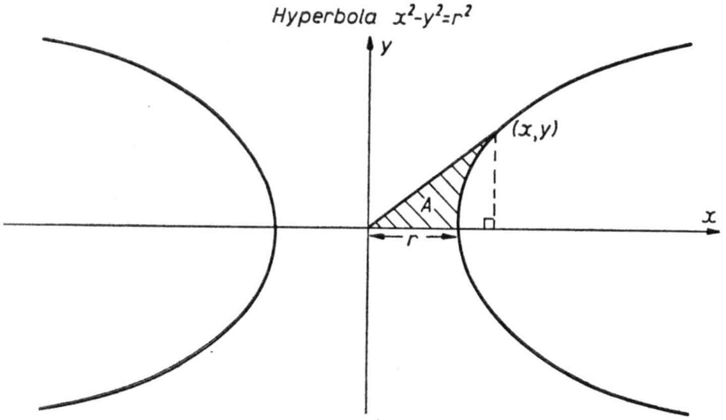

FIG. 5

This is analogous to defining $\sin\theta$ and $\cos\theta$ using the circle $x^2+y^2 = r^2$. See Fig. 6.

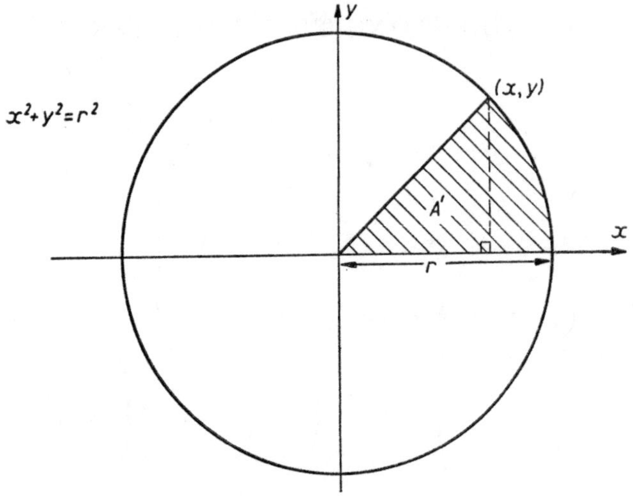

$$\cos\theta' = \frac{x}{r}, \quad \sin\theta' = \frac{y}{r}, \quad \theta' = \frac{2A'}{r^2}$$

FIG. 6

It is interesting to notice that the matrix

$$\mathbf{P} = \begin{pmatrix} 1 & 0 \\ 0 & i \end{pmatrix}$$

transforms the circle $x^2+y^2 = r^2$ into the hyperbola $X^2-Y^2 = r^2$. Since $|\mathbf{P}| = i$, the area of transformation is i, so that $iA' = A$, which means that $i\theta' = \theta$. That is,

$$i\theta \text{ radians} = \theta \text{ hyperbolic radians}$$

By considering the transformation

$$\begin{pmatrix} 1 & 0 \\ 0 & i \end{pmatrix} \begin{pmatrix} \cos\theta' \\ \sin\theta' \end{pmatrix} = \begin{pmatrix} \cos\theta' \\ i\sin\theta' \end{pmatrix} = \begin{pmatrix} \cosh\theta \\ \sinh\theta \end{pmatrix}$$

we have shown that $\cos(-i\theta) = \cos i\theta = \cosh\theta$ and $i\sin(-i\theta) = -i\sin i\theta = \sinh\theta$. Thus these relationships have been established without recourse to the exponential function. By showing that A (see Fig. 5) is given by

$$A = \int_0^y (r^2+y^2)^{\frac{1}{2}}\,dy - \tfrac{1}{2}xy = \frac{r^2}{2}\log_e\left(\frac{x+y}{r}\right)$$

it is easy to show that $\cosh\theta = \frac{1}{2}(e^{\theta}+e^{-\theta})$ and $\sinh\theta = \frac{1}{2}(e^{\theta}-e^{-\theta})$.

1.6. If $\mathbf{X}$ is a latent vector of $\mathbf{AB}$ and $\mathbf{Y}$ the corresponding vector of $\mathbf{BA}$, then $\mathbf{BX} = \mathbf{Y}$ (and $\mathbf{AY} = \mathbf{X}$).

1.7. The matrix of exercise 1.1 (iv) has the same characteristic equation as $\mathbf{I}$.

1.10.
$$\mathbf{AX} = \lambda\mathbf{X}$$
Hence
$$\mathbf{X}^*\mathbf{A}^* = \bar{\lambda}\mathbf{X}^* \quad (\bar{\lambda} \text{ is the complex conjugate of } \lambda)$$
which gives
$$\mathbf{X}^*\mathbf{A}^*\mathbf{AX} = \bar{\lambda}\mathbf{X}^*\lambda\mathbf{X} = \bar{\lambda}\lambda\mathbf{X}^*\mathbf{X}$$
But
$$\mathbf{A}^*\mathbf{A} = \mathbf{I}$$
so that
$$\mathbf{X}^*\mathbf{X} = \bar{\lambda}\lambda\mathbf{X}^*\mathbf{X}$$
Hence
$$\bar{\lambda}\lambda = 1 \quad \text{and} \quad |\lambda| = 1$$

1.13. See theorem 4.3 for the first part.
From theorem 1.14,
$$h(\mathbf{A})\,\mathbf{X} = h(\lambda)\,\mathbf{X}$$
But
$$h(\mathbf{A}) = 0$$
Hence
$$h(\lambda) = 0$$

1.18. (i)
$$\lambda_1 = 15, \quad \lambda_2, \lambda_3 = \pm 2\sqrt{6}$$
(ii)
$$\lambda_1 = 34, \quad \lambda_2, \lambda_3 = \pm 4\sqrt{5}, \quad \lambda_4 = 0$$

1.20. When $\lambda = -a$, $\mathbf{X}^T = \left(-\sum_{i=2}^{n} x_i \quad x_2 \quad x_3 \quad \dots \quad x_n\right)$.

When $\lambda = a(n-1), \mathbf{X}^T = k(11\dots1)$.

1.28. For
$$\mathbf{P} = \begin{pmatrix} \mathbf{A} & \lambda\mathbf{I} \\ -\mathbf{I} & 0 \end{pmatrix} \quad \text{and} \quad \mathbf{Q} = \begin{pmatrix} \mathbf{B} & \lambda\mathbf{I} \\ -\mathbf{I} & -\mathbf{A} \end{pmatrix}$$
form $\mathbf{PQ}$ and $\mathbf{QP}$.
$$|\mathbf{PQ}| = |\mathbf{QP}|$$

Chapter 2

2.1. Axes of symmetry are the lines
$$4y = 3x \quad \text{and} \quad 3y = -4x$$
Distances are $\frac{1}{7}$ and i. (Since the curve is a hyperbola only one axis actually meets the curve.)

2.2. One axis of symmetry is the line $48x = 28y = 21z$. Any line in the plane $x+2y+2z = 0$ is also an axis of symmetry. (One cross-section of the ellipsoid is a circle.)

The distances are $\frac{1}{8}$ and 1. (The radius of the circle is 1.)

2.3. Ellipse if $\lambda_1, \lambda_2 > 0$ or $\lambda_1, \lambda_2 < 0$.
Circle if $\lambda_1 = \lambda_2 \neq 0$.
Hyperbola if $\lambda_1 > 0, \lambda_2 < 0$ or $\lambda_1 < 0, \lambda_2 > 0$.
Parabola if $\lambda_1 = 0$ or $\lambda_2 = 0$.
(Straight line if $\lambda_1 = \lambda_2 = 0$.)

2.4. (i) $\lambda_1 = 2, \lambda_3 = -3$, hence hyperbola.
(ii) $\lambda_1 = 13, \lambda_2 = 0$, hence parabola.

2.6. Jacobi scheme is

$$\begin{aligned}
x_{1,r+1} &= \tfrac{1}{2}(3 - 3x_{3r} - 3x_{4r})\\
x_{2,r+1} &= (7 - 2x_{1r} + 0{\cdot}1x_{3r} + 0{\cdot}1x_{4r})\\
x_{3,r+1} &= \tfrac{1}{200}(-39 + 40x_{1r} + 20x_{2r} + 2x_{4r})\\
x_{4,r+1} &= \tfrac{1}{100}(19{\cdot}5 - 20x_{1r} - 10x_{2r} + x_{3r})
\end{aligned}$$

with $x_{10} = x_{20} = x_{30} = x_{40} = 0$.

The Gauss–Seidel scheme is

$$\begin{aligned}
x_{1,r+1} &= \tfrac{1}{2}(3 - 3x_{3r} - 3x_{4r})\\
x_{2,r+1} &= (7 - 2x_{1,r+1} + 0{\cdot}1x_{3r} + 0{\cdot}1x_{4r})\\
x_{3,r+1} &= \tfrac{1}{200}(-39 + 40x_{1,r+1} + 20x_{2,r+1} + 2x_{4r})\\
x_{4,r+1} &= \tfrac{1}{100}(19{\cdot}5 - 20x_{1,r+1} - 10x_{2,r+1} + x_{3,r+1})
\end{aligned}$$

with $x_{30} = x_{40} = 0$ being the only initial values needed. The exact solution is $x_1 = 1{\cdot}5, x_2 = 4{\cdot}0, x_3 = 0{\cdot}5, x_4 = -0{\cdot}5$.

2.7. (i) For Jacobi's method, $|\mathbf{M} - \lambda\mathbf{I}| = -\lambda^3 + 0{\cdot}01 = 0$. Hence the method will converge.

For the Gauss–Seidel method, $|\mathbf{M} - \lambda\mathbf{I}| = -\lambda(\lambda^2 + 3{\cdot}98\lambda - 3{\cdot}99) = 0$. Hence the method will not converge.

(ii) For Jacobi's method, $|\mathbf{M} - \lambda\mathbf{I}| = (\lambda - 2)(-\lambda^2 - 2\lambda + 4{\cdot}1) = 0$. Hence the method will not converge.

For the Gauss–Seidel method, $|\mathbf{M} - \lambda\mathbf{I}| = \lambda^2(-\lambda - 0{\cdot}1) = 0$. Hence the method will converge.

Solution of (iii) is $x = y = z = 1$.
Solution of (iv) is $x = y = z = 1$.

2.12. (i)
$$\mathbf{X} = k_1 e^{5t}\begin{pmatrix} 1 \\ 1 \end{pmatrix} + k_2 e^{3t}\begin{pmatrix} 1 \\ -1 \end{pmatrix}$$

(ii)
$$\mathbf{X} = k_1 e^{(1+i)t}\begin{pmatrix} 1 \\ -2+i \end{pmatrix} + k_2 e^{(1-i)t}\begin{pmatrix} 1 \\ -2-i \end{pmatrix}$$

(iii)
$$\mathbf{X} = 3e^{4t}\begin{pmatrix} 1 \\ 1 \end{pmatrix} + 2e^{-2t}\begin{pmatrix} 1 \\ 7 \end{pmatrix}$$

(iv)
$$\mathbf{X} = \begin{pmatrix} k_1 - 2k_2 - 10k_2 t \\ 2k_1 + k_2 - 20k_2 t \end{pmatrix} e^{-3t}$$

(v)
$$\mathbf{X} = \frac{e^{2t}}{\sqrt{2}}\begin{pmatrix} 1-2t \\ 1+2t \end{pmatrix}$$

(vi)
$$\mathbf{X} = 3e^{2t}\begin{pmatrix} 1 \\ -1 \end{pmatrix} - 2e^{3t}\begin{pmatrix} 1 \\ -2 \end{pmatrix}$$

(vii)
$$\mathbf{X} = k_1 e^{t}\begin{pmatrix} 9 \\ 1 \\ 7 \end{pmatrix} + k_2 e^{-t}\begin{pmatrix} -3 \\ 1 \\ -3 \end{pmatrix} + k_3\begin{pmatrix} 1 \\ 0 \\ 1 \end{pmatrix}$$

2.13. (i) The latent roots of $\mathbf{A}_2$ are $\lambda_1 = \lambda_2 = 1$, since these are both roots of $\mathbf{A}$. The latent vector of $\mathbf{A}_2$ is

$$\mathbf{X} = k\begin{pmatrix} 3 \\ -4 \end{pmatrix}$$

Hence

$$\mathbf{C}_2 = \begin{pmatrix} 1 & 0 & 0 \\ 0 & \frac{3}{5} & \frac{4}{5} \\ 0 & -\frac{4}{5} & \frac{3}{5} \end{pmatrix}$$

(ii)
$$\mathbf{X} = \begin{pmatrix} x_1 \\ x_2 \\ x_3 \end{pmatrix} = \frac{k_1}{3}\begin{pmatrix} 1 \\ 2 \\ 2 \end{pmatrix} e^{2t} + \frac{1}{15}\begin{pmatrix} -27k_2 - 1836k_3 - 2025k_3 t \\ -60k_2 - 3705k_3 - 4500k_3 t \\ -39k_2 - 3702k_3 - 2925k_3 t \end{pmatrix} e^{t}$$

(iii) (a)
$$\begin{pmatrix} x_1 \\ x_2 \\ x_3 \end{pmatrix} = \frac{k_1}{\sqrt{2}}\begin{pmatrix} 1 \\ 0 \\ -1 \end{pmatrix} e^{3t} + \frac{1}{\sqrt{6}}\begin{pmatrix} -10k_2 - 8\sqrt{(2)}\,k_3 - 15\sqrt{(2)}\,k_3 t \\ 2k_2 + \sqrt{(2)}\,k_3 + 3\sqrt{(2)}\,k_3 t \\ 8k_2 + 10\sqrt{(2)}\,k_3 + 12\sqrt{(2)}\,k_3 t \end{pmatrix} e^{2t}$$

(b)
$$\begin{pmatrix} x_1 \\ x_2 \\ x_3 \end{pmatrix} = \frac{1}{3}\begin{pmatrix} k_1 + 2k_2 + 2k_3 + 9(k_2 + k_3)t + \frac{81}{2}k_3 t^2 \\ 2k_1 - 2k_2 + k_3 + 18(k_2 - 2k_3)t + 81k_3 t^2 \\ 2k_1 + k_2 - 2k_3 + 9(2k_2 - k_3)t + 81k_3 t^2 \end{pmatrix} e^{t}$$

2.16. (i) $$\begin{pmatrix} x_n \\ y_n \end{pmatrix} = \frac{1}{6}\begin{pmatrix} 1 \\ 1 \end{pmatrix} 3^n + \frac{1}{2}\begin{pmatrix} 1 \\ -1 \end{pmatrix}(-1)^n$$

(ii) $$\begin{pmatrix} x_n \\ y_n \end{pmatrix} = \frac{1}{\sqrt{2}}\begin{pmatrix} k_1 - k_2 + 4k_2 n \\ k_1 + k_2 + 4k_2 n \end{pmatrix}(-1)^n$$

Chapter 3

3.1. (i) $$\lambda_1 = 3, \quad \lambda_2 = 2, \quad \lambda_3 = -4.$$

$$\mathbf{X}_1 = k\begin{pmatrix} 3 \\ 6 \\ 2 \end{pmatrix}, \quad \mathbf{X}_2 = k\begin{pmatrix} 1 \\ 3 \\ 1 \end{pmatrix}, \quad \mathbf{X}_3 = k\begin{pmatrix} 19 \\ 45 \\ 1 \end{pmatrix}$$

(ii) $$\lambda_1 = \lambda_2 = \lambda_3 = 1, \quad \mathbf{X} = k\begin{pmatrix} -4 \\ 2 \\ 1 \end{pmatrix}$$

(iii) $$\lambda_1 = \lambda_2 = 2, \quad \lambda_3 = \lambda_4 = \lambda_5 = 1$$

$$\mathbf{X}_1 = k\begin{pmatrix} 41 \\ -15 \\ -7 \\ 10 \\ 1 \end{pmatrix} \qquad \mathbf{X}_3 = k\begin{pmatrix} -4 \\ 2 \\ 1 \\ -3 \\ 0 \end{pmatrix}$$

3.2. Using arithmetic correct to four significant figures gives

$$\mathbf{A}_1 = \begin{pmatrix} -1 & 100 & -670 \\ 0{\cdot}05 & 6{\cdot}95 & -6 \\ 0 & 1 & 0 \end{pmatrix}$$

Notice that in exact arithmetic the a_{21} position is zero and hence $\mathbf{A}_2 = \mathbf{B}$ is not formed. This example highlights the dangers of a small pivot element. The roots of

$$\lambda^3 - 5{\cdot}95\lambda^2 - 5{\cdot}95\lambda + 39{\cdot}5 = 0$$

are

$$\lambda_1 = 5{\cdot}80, \quad \lambda_2 = 2{\cdot}68, \quad \lambda_3 = -2{\cdot}54$$

which do not bear much resemblance to the correct latent roots. Notice that b_1 is quite close to the trace of $\mathbf{A}$, but that b_3 is quite different from $|\mathbf{A}|$. Most computers use floating point arithmetic, which means that calculations are performed to a set number of significant figures, so that this sort of result is quite possible in practice.

3.3.

$$\mathbf{X} = k\begin{pmatrix} 1 \\ 1 \\ 2 \end{pmatrix}$$

Chapter 4

4.2. This starting vector yields the full characteristic equation

$$\lambda^4 - 3\lambda^3 - 7\lambda^2 - 13\lambda - 10 = 0$$

The grade of this vector with respect to **A** is four.

4.3. From §4.2 we have $\mathbf{F} = \mathbf{Y}^{-1}\mathbf{A}\mathbf{Y}$, where, of course,

$$\mathbf{Y} = (\, \mathbf{Y}_0 \quad \mathbf{Y}_1 \quad \dots \quad \mathbf{Y}_{r-1}\,)$$

A latent vector, say **Z**, of **F** is easily found and then

$$\mathbf{X} = \mathbf{YZ}$$

where **X** is the corresponding vector of **A**. If $r < n$, then $\mathbf{Y}^{-1}$ does not exist so that **F** is not similar to **A**. (It is not even the same size.)

Chapter 5

5.1. (i)

$$f_2(x) = 27x^2 - 40$$

$$f_1(x) = \tfrac{1}{3}(80x + 63)$$

$$f_0(x) \text{ is positive}$$

The exact roots are $x_1 = \frac{7}{3}$; $x_2, x_3 = \frac{1}{6}(-7 \pm \sqrt{13})$

(ii)

$$f_3(x) = 4x^3 + 12x^2 + 12x + 4$$

$$f_2(x) = 25$$

The two real roots are $x_1, x_2 = -1 \pm \sqrt{5}$.

Notice that having only three polynomials in the sequence means that at most only two distinct real roots are possible because there cannot be more than two changes of sign.

5.4. See Lanczos† for an excellent discussion on orthogonal polynomials.

5.5. (i) Exact roots are $\lambda_1 = -3$; $\lambda_2, \lambda_3 = -3 \pm \sqrt{8}$.

(ii) Exact roots are $\lambda_1 = -1$; $\lambda_2, \lambda_3 = 1 \pm \sqrt{3}$.

5.7. (i) It is not known under what conditions convergence takes place. The advantage is that complex arithmetic is avoided.

(ii) The exact roots are

$$\lambda_3, \lambda_4 = \pm ti; \quad \lambda_1, \lambda_2 = \pm \frac{1}{t} i$$

where t is the 'golden ratio', i.e.

$$t = \frac{1 + \sqrt{5}}{2}$$

† Reference 18.

Chapter 6

6.2. (i)

$$\mathbf{B} = \begin{pmatrix} 6 & 5 & 0 \\ 5 & 9{\cdot}2 & -1{\cdot}6 \\ 0 & -1{\cdot}6 & 1{\cdot}8 \end{pmatrix}, \quad \lambda_1 = 13, \quad \lambda_2 = 3, \quad \lambda_3 = 1$$

$$\mathbf{X}_1 = \begin{pmatrix} 1 \\ 1 \\ 1 \end{pmatrix}, \quad \mathbf{X}_2 = \begin{pmatrix} 1 \\ -1 \\ 0 \end{pmatrix}, \quad \mathbf{X}_3 = \begin{pmatrix} 1 \\ 1 \\ -2 \end{pmatrix}$$

(ii)

$$\mathbf{B} = \begin{pmatrix} 1 & 10 & 0 \\ 10 & -20 & 0 \\ 0 & 0 & 5 \end{pmatrix}, \quad \lambda_1 = \lambda_2 = 5, \quad \lambda_3 = -24$$

$$\mathbf{X}_1, \mathbf{X}_2 = \begin{pmatrix} 1 \\ \frac{8}{25} + \frac{3k}{5} \\ -\frac{6}{25} + \frac{4k}{5} \end{pmatrix}, \quad \mathbf{X}_3 = \begin{pmatrix} 1 \\ -2 \\ \frac{3}{2} \end{pmatrix}$$

From the above set of vectors, any two orthogonal vectors can be chosen for $\mathbf{X}_1$ and $\mathbf{X}_2$. Notice $\mathbf{X}_3$ is orthogonal to $\mathbf{X}_1, \mathbf{X}_2$ for all values of k.

6.3.

$$\mathbf{B} = \begin{pmatrix} 2 & 5 & 0 & 0 \\ 5 & -2 & 5 & 0 \\ 0 & 5 & 6 & -2 \\ 0 & 0 & -2 & 3 \end{pmatrix}$$

To the nearest integer the roots are $\lambda_1 = 9$, $\lambda_2 = 5$, $\lambda_3 = 2$, $\lambda_4 = -7$. Correct to two decimal places $\lambda_3 = 2{\cdot}00$ and $\lambda_4 = -6{\cdot}84$.

$$\mathbf{Y}_3 = \begin{pmatrix} 1 \\ 0 \\ -1 \\ -2 \end{pmatrix}, \quad \mathbf{X}_3 = \begin{pmatrix} 1 \\ -\frac{3}{5} \\ \frac{4}{5} \\ -2 \end{pmatrix}; \quad \mathbf{Y}_4 \simeq \begin{pmatrix} 1{\cdot}00 \\ -1{\cdot}77 \\ 0{\cdot}71 \\ 0{\cdot}15 \end{pmatrix}, \quad \mathbf{X}_4 \simeq \begin{pmatrix} 1{\cdot}00 \\ -1{\cdot}60 \\ -0{\cdot}82 \\ 0{\cdot}66 \end{pmatrix}$$

Chapter 7

7.1. (i)

$$\mathbf{B} = \begin{pmatrix} 1 & -25 & 0 \\ -25 & 5 & 12 \\ 0 & 12 & -2 \end{pmatrix}$$

(ii)

$$\mathbf{B} = \begin{pmatrix} 5 & -9 & 0 & 0 \\ -9 & 9 & 5 & 0 \\ 0 & 5 & 2{\cdot}4 & -1{\cdot}2 \\ 0 & 0 & -1{\cdot}2 & -9{\cdot}4 \end{pmatrix}$$

Chapter 8

8.1. (i)

$$\mathbf{C} = \begin{pmatrix} 6 & 25 & 0 \\ 1 & 9{\cdot}2 & 2{\cdot}56 \\ 0 & 1 & 1{\cdot}8 \end{pmatrix}, \quad \mathbf{Y} = \begin{pmatrix} 1 & 0 & 0 \\ 0 & 3 & 6{\cdot}4 \\ 0 & 4 & -4{\cdot}8 \end{pmatrix}$$

(ii)

$$\mathbf{C} = \begin{pmatrix} 2 & 25 & 0 & 0 \\ 1 & -2 & 25 & 0 \\ 0 & 1 & 6 & 4 \\ 0 & 0 & 1 & 3 \end{pmatrix}, \quad \mathbf{Y} = \begin{pmatrix} 1 & 0 & 0 & 0 \\ 0 & 4 & -9 & -24 \\ 0 & 3 & 12 & 32 \\ 0 & 0 & 20 & -30 \end{pmatrix}$$

In both the above exercises it is assumed that $\mathbf{Y}_1^T = (\,1 \quad 0 \quad \ldots \quad 0\,)$.

(iii)

$$\mathbf{C} = \begin{pmatrix} -1 & 9 & 0 & 0 \\ 1 & -\frac{77}{9} & \frac{200}{81} & 0 \\ 0 & 1 & \frac{50}{9} & 0 \\ 0 & 0 & 0 & 0 \end{pmatrix}$$

Note that $\mathbf{Y}_4 = 0$.

8.3. (i)

$$\mathbf{C} = \begin{pmatrix} 4 & 6 & 0 \\ 1 & -3 & -24 \\ 0 & 1 & 5 \end{pmatrix}, \quad \mathbf{Y} = \begin{pmatrix} 1 & 0 & 0 \\ 0 & 1 & -2 \\ 0 & -1 & 3 \end{pmatrix} \quad \text{if} \quad \mathbf{Y}_1 = \begin{pmatrix} 1 \\ 0 \\ 0 \end{pmatrix}$$

$$\lambda_1 = 1, \lambda_2 = 2, \lambda_3 = 3, \quad \mathbf{X}_1 = \begin{pmatrix} 1 \\ -\frac{1}{3} \\ \frac{5}{12} \end{pmatrix}, \quad \mathbf{X}_2 = \begin{pmatrix} 1 \\ -\frac{5}{9} \\ \frac{2}{3} \end{pmatrix}, \quad \mathbf{X}_3 = \begin{pmatrix} 1 \\ -\frac{3}{4} \\ \frac{7}{8} \end{pmatrix}$$

Vectors of $\mathbf{A}^T$ are

$$\mathbf{W}_1 = \begin{pmatrix} 1 \\ -15 \\ -14 \end{pmatrix}, \quad \mathbf{W}_2 = \begin{pmatrix} 1 \\ -22 \\ -20 \end{pmatrix}, \quad \mathbf{W}_3 = \begin{pmatrix} 1 \\ -27 \\ -24 \end{pmatrix}$$

(ii)

$$\mathbf{C} = \begin{pmatrix} 4 & -2 & 0 \\ 1 & 2 & b_3 \\ 0 & 0 & 1 \end{pmatrix} \quad \text{if} \quad \mathbf{Y}_1 = \begin{pmatrix} 1 \\ 0 \\ 0 \end{pmatrix}$$

Note that $\mathbf{Y}_3 = 0$, hence b_3 will depend on the next vector chosen.

$$\lambda_1 = 1, \quad \lambda_2, \lambda_3 = 3 \pm i, \quad \mathbf{X}_1 = \begin{pmatrix} 1 \\ \frac{29}{6} \\ \frac{4}{3} \end{pmatrix}, \quad \mathbf{X}_2, \mathbf{X}_3 = \begin{pmatrix} 1 \\ \frac{1 \mp i}{2} \\ 1 \mp i \end{pmatrix}$$

Vectors of $\mathbf{A}^T$ are

$$\mathbf{W}_1 = \begin{pmatrix} 0 \\ 1{\cdot}2 \\ -0{\cdot}6 \end{pmatrix}$$

Notice that

$$\mathbf{C}_1 = \begin{pmatrix} 4 & -2 & 0 \\ 1 & 2 & 0 \\ 0 & 1 & 1 \end{pmatrix}$$

and when $\lambda = 1$, $y_1 \neq 1$. (See § 6.3.)

$$\mathbf{W}_1, \mathbf{W}_2 = \begin{pmatrix} 1 \\ -\frac{2(1 \pm 2i)}{25} \\ -\frac{23 \pm 29i}{50} \end{pmatrix}$$

(iii)

$$\mathbf{C} = \begin{pmatrix} 1 & 3 & 0 \\ 1 & -1 & 0 \\ 0 & 0 & 2 \end{pmatrix} \quad \text{if } \mathbf{Y}_1 = \begin{pmatrix} 1 \\ 0 \\ 0 \end{pmatrix}$$

Note that $\mathbf{Y}_3 = \mathbf{Z}_3 = 0$ so that $b_3 = 0$.

$$\lambda_1 = \lambda_2 = 2, \quad \lambda_3 = -2, \quad \mathbf{X}_1, \mathbf{X}_2 = \begin{pmatrix} 1 \\ k + \frac{5}{3} \\ k + \frac{4}{3} \end{pmatrix}, \quad \mathbf{X}_3 = \begin{pmatrix} 1 \\ -5 \\ -4 \end{pmatrix}$$

Vectors of $\mathbf{A}^T$ are

$$\mathbf{W}_1, \mathbf{W}_2 = \begin{pmatrix} 1 \\ 1 - 4k \\ -1 + 5k \end{pmatrix}, \quad \mathbf{W}_3 = \begin{pmatrix} 1 \\ -3 \\ 3 \end{pmatrix}$$

(iv)

$$\lambda_1 = \lambda_2 = 4, \quad \lambda_3 = -2, \quad \mathbf{X}_1, \mathbf{X}_2 = \begin{pmatrix} 1 \\ -1 \\ -5 \end{pmatrix}, \quad \mathbf{X}_3 = \begin{pmatrix} 1 \\ -1 \\ 1 \end{pmatrix}$$

Vectors of $\mathbf{A}^T$ are

$$\mathbf{W}_1, \mathbf{W}_2 = \begin{pmatrix} 1 \\ 1 \\ 0 \end{pmatrix}, \quad \mathbf{W}_3 = \begin{pmatrix} 1 \\ -\frac{7}{3} \\ \frac{2}{3} \end{pmatrix}$$

Note that if $\mathbf{Y}_1^T = \mathbf{Z}_1^T = (\,1 \quad 0 \quad 0\,)$ then $\mathbf{Z}_2^T \mathbf{Y}_2 = 0$.

Chapter 9

9.1. (i)

$$\lambda = 9, \quad \mathbf{X} = \begin{pmatrix} 1 \\ \frac{1}{2} \\ -1 \end{pmatrix}$$

(ii)

$$\lambda_1 = -\lambda_2 = 15, \quad \mathbf{X}_1 = \begin{pmatrix} 143 \\ 45 \\ -67 \end{pmatrix}, \quad \mathbf{X}_2 = \begin{pmatrix} 1 \\ 0 \\ 1 \end{pmatrix}$$

(iii)

$$\lambda_1, \lambda_2 = 10 \pm 5i, \quad \mathbf{X}_1, \mathbf{X}_2 = \begin{pmatrix} 7 \mp 12i \\ 9 \pm 5i \\ 2 \mp 3i \end{pmatrix}$$

9.2. $x = 9{\cdot}40$.

Iterating with the Frobenius matrix is equivalent to using Bernoulli's method for finding the root of the largest modulus of a polynomial.

9.3. (ii) Exact roots are $\lambda_1, \lambda_2 = \pm 4\sqrt{(5)}\,i$.

9.5. $\lambda_1 = 11$, $\lambda_3 = 3 - \sqrt{5}$.

9.6. The equations $\mathbf{LZ} = k_0 \mathbf{Y}_0$, which are easily solved by forward substitution, give

$$z_1 = 0{\cdot}1, \quad z_2 = 0{\cdot}2, \quad z_3 = 0{\cdot}5$$

Then $\mathbf{UY}_1 = \mathbf{Z}$ can be solved by backward substitution to give

$$y_1 = -2{\cdot}2, \quad y_2 = -1{\cdot}3 \quad y_3 = 0{\cdot}5$$

Hence

$$\mathbf{Y}_1 = -2{\cdot}2 \begin{pmatrix} 1{\cdot}000 \\ 0{\cdot}591 \\ -0{\cdot}227 \end{pmatrix}, \quad k_1^{-1} = -2{\cdot}2$$

Solving $\mathbf{LUY}_2 = k_1 \mathbf{Y}_1$ gives

$$\mathbf{Y}_2 = -3{\cdot}082 \begin{pmatrix} 1{\cdot}000 \\ 0{\cdot}578 \\ -0{\cdot}227 \end{pmatrix}, \quad k_2^{-1} = -3{\cdot}082$$

Solving $\mathbf{LUY}_3 = k_2 \mathbf{Y}_2$ gives

$$\mathbf{Y}_3 = -3{\cdot}065 \begin{pmatrix} 1{\cdot}000 \\ 0{\cdot}578 \\ -0{\cdot}227 \end{pmatrix}$$

Hence

$$\lambda = \frac{1}{-3{\cdot}065} + 10 = 9{\cdot}67$$

9.7. $\mathbf{A}_1 - 4{\cdot}7150\mathbf{I}$

$$= \begin{pmatrix} -4{\cdot}7150 & 5{\cdot}0000 & 0{\cdot}0000 \\ 5{\cdot}0000 & -5{\cdot}3150 & -0{\cdot}20000 \\ 0{\cdot}0000 & -0{\cdot}2000 & -3{\cdot}1150 \end{pmatrix}$$

$$= \begin{pmatrix} 1 & 0 & 0 \\ -1{\cdot}06045 & 1 & 0 \\ 0{\cdot}00000 & 15{\cdot}6863 & 1 \end{pmatrix} \begin{pmatrix} -4{\cdot}71500 & 5{\cdot}00000 & 0{\cdot}00000 \\ 0 & -0{\cdot}01275 & -0{\cdot}20000 \\ 0 & 0 & 0{\cdot}02226 \end{pmatrix} = \mathbf{LU}$$

Then

$$\begin{pmatrix} -4{\cdot}71500 & 5{\cdot}00000 & 0 \\ 0 & -0{\cdot}01275 & -0{\cdot}20000 \\ 0 & 0 & 0{\cdot}02226 \end{pmatrix} \begin{pmatrix} y_1 \\ y_2 \\ y_3 \end{pmatrix} = \begin{pmatrix} 1 \\ 1 \\ 1 \end{pmatrix}$$

gives

$$\mathbf{Y}_1 = \begin{pmatrix} -7556{\cdot}1774 \\ -7125{\cdot}2753 \\ 44{\cdot}92363 \end{pmatrix} = -7556{\cdot}1774 \begin{pmatrix} 1 \\ 0{\cdot}9430 \\ -0{\cdot}0595 \end{pmatrix} k_1^{-1} = -7556{\cdot}1774$$

Solving $\mathbf{LUY}_2 = k_1 \mathbf{Y}_1$ gives

$$\mathbf{Y}_2 = \begin{pmatrix} 23362{\cdot}142 \\ 22030{\cdot}700 \\ -1414{\cdot}4744 \end{pmatrix} = 2336{\cdot}142 \begin{pmatrix} 1 \\ 0{\cdot}9430 \\ -0{\cdot}0605 \end{pmatrix}$$

This gives, correct to four decimal places, the latent vector of $\mathbf{A}_1$. (The latent root was only given to four decimal places.) Notice that $\mathbf{Y}_1$ is extremely accurate.

9.8.

$$\lambda_2 = \lambda_3 = 1, \quad \mathbf{X}_2 = \mathbf{X}_3 = \begin{pmatrix} 9 \\ 20 \\ 13 \end{pmatrix}$$

(Compare with exercise 2.13.)

9.9.

$$\mathbf{C}_1 = \begin{pmatrix} \frac{1}{2} & \frac{1}{2} & \frac{1}{2} & \frac{1}{2} \\ \frac{1}{2} & \frac{1}{2} & -\frac{1}{2} & -\frac{1}{2} \\ \frac{1}{2} & -\frac{1}{2} & \frac{1}{2} & -\frac{1}{2} \\ \frac{1}{2} & -\frac{1}{2} & -\frac{1}{2} & \frac{1}{2} \end{pmatrix}$$

which gives

$$\mathbf{C}_1^{-1}\mathbf{A}\mathbf{C}_1 = \begin{pmatrix} 12 & 3 & 2 & -2 \\ 0 & 1 & -1 & -1 \\ 0 & 5 & 5 & -5 \\ 0 & -5 & -6 & 4 \end{pmatrix}$$

so

$$\mathbf{A}_3 = \begin{pmatrix} 1 & -1 & -1 \\ 5 & 5 & -5 \\ -5 & -6 & 4 \end{pmatrix}$$

Iteration gives

$$\lambda_2 = 10 \quad \text{and} \quad \mathbf{Z}_2 = \begin{pmatrix} 0 \\ 1 \\ -1 \end{pmatrix}$$

which gives

$$\mathbf{X}_2 = \begin{pmatrix} \frac{1}{2} \\ \frac{1}{2} \\ 0 \\ 1 \end{pmatrix}$$

$$\mathbf{C}_2 = \begin{pmatrix} 0 & \frac{1}{\sqrt{2}} & -\frac{1}{\sqrt{2}} \\ \frac{1}{\sqrt{2}} & \frac{1}{2} & \frac{1}{2} \\ -\frac{1}{\sqrt{2}} & \frac{1}{2} & \frac{1}{2} \end{pmatrix}$$

which gives

$$\mathbf{C}_2^{-1}\,\mathbf{A}_3\,\mathbf{C}_2 = \begin{pmatrix} 10 & 5+\frac{1}{\sqrt{2}} & -5+\frac{1}{\sqrt{2}} \\ 0 & -\frac{1}{\sqrt{2}} & -1-\frac{1}{\sqrt{2}} \\ 0 & -1+\frac{1}{\sqrt{2}} & \frac{1}{\sqrt{2}} \end{pmatrix}$$

which gives

$$\lambda_3 = 1, \quad \lambda_4 = -1$$

and

$$\mathbf{X}_3 = \begin{pmatrix} 23 \\ 23 \\ -54 \\ 1 \end{pmatrix}, \quad \mathbf{X}_4 = \begin{pmatrix} 105 \\ -38 \\ -77 \\ 1 \end{pmatrix}$$

Chapter 10

10.1. (i) $\lambda_1 = 20$, $\lambda_2 = 1$.
(ii) $\lambda_1 = -\lambda_2 = 5$.
(iii) $\lambda_1 = 37$, $\lambda_2 = 6$, $\lambda_3 = 1$.
(iv) $\lambda_1 = 35$; $\lambda_2, \lambda_3 = \pm i$.

BIBLIOGRAPHY

REFERENCES CITED

1. VARGA, R. S., *Matrix Iterative Analysis* (Prentice-Hall, New Jersey, 1962).
2. WILKES, M. V., *A Short Introduction to Numerical Analysis* (Cambridge University Press, Cambridge, 1966).
3. FOX, L., *Introduction to Numerical Linear Algebra* (Oxford University Press, London, 1964).
4. SMITH, G. D., *Numerical Solution of Partial Differential Equations* (Oxford University Press, London, 1965).
5. NOBLE, B., *Numerical Methods* (Oliver and Boyd, Edinburgh, 1966).
6. FADEEVA, V. N., *Computational Methods of Linear Algebra* (Dover Publications, New York, 1959).
7. WILKINSON, J. H., *The Algebraic Eigenvalue Problem* (Oxford University Press, London, 1965).
8. BEREZIN, I. S., and ZHIDKOV, N. P., *Computing Methods* (Pergamon Press, Oxford, 1965).
9. MULLER, D. E., *A Method for Solving Algebraic Equations Using an Automatic Computer* (Mathematical Tables, Washington, **10**, 208–215, 1956).
10. GANTMACHER, F. R., *Matrix Theory* (Chelsea, New York, 1960).
11. ARCHBOLD, J. W., *Algebra* (Pitman, London, 1961).
12. HOUSEHOLDER, A. S., *The Theory of Matrices in Numerical Analysis* (Blaisdell, New York, 1964).
13. WALSH, J. (Ed.), *Numerical Analysis: An Introduction* (Academic Press, London, 1966).
14. KAISER, H. F., 'A method for determining eigenvalues', *J. Soc. app. Maths.*, 1964, **12** (No. 1), 238–247.
15. JENNINGS, W., *First Course in Numerical Analysis* (Macmillan, New York, 1964).
16. FOX, L., and PARKER, I. B., *Chebyshev Polynomials in Numerical Analysis* (Oxford University Press, London, 1968).
17. WILKINSON, J. H., *Rounding Errors in Algebraic Processes* (H.M.S.O., London, 1963).
18. LANCZOS, C., *Applied Analysis* (Pitman, London, 1967).
19. FRANKLIN, J. N., *Matrix Theory* (Prentice-Hall, New Jersey, 1968).
20. BISHOP, R. E. D., GLADWELL, G. M. L., and MICHAELSON, S., *Matrix Analysis of Vibration* (Cambridge University Press, Cambridge, 1965).
21. FADEEV, D. K., and FADEEVA, V. N., *Computational Methods of Linear Algebra* (W. H. Freeman, San Francisco, 1963).
22. EBERLEIN, P. J., 'A Jacobi-like method for the automatic computation of eigenvalues and eigenvectors of an arbitrary matrix', *J. Soc. indust. appl. Maths.*, 1962, **10**, 74–88.
23. FORSYTHE, G. E., and MOLER, C. B., *Computer Solution of Linear Algebraic Systems* (Prentice-Hall, New Jersey, 1967).

ADDITIONAL REFERENCES

BELLMAN, R., *Introduction to Matrix Analysis* (McGraw-Hill, New York, 1960).

DANILEVSKY, A., 'O čislennom rešĕnii vekovogo uravnenija', *Mat. Sb.*, 1937, **44** (No. 2), 169–171.

FLETCHER, R., 'A technique for orthogonalization', *J. Inst. Maths Applics.*, 1969, **5**, 162–166.

FRANCIS, J. G. F., 'The **QR** transformation', Parts I and II, *Computer J.*, 1961, **4**, 265–271; 1962, **4**, 332–345.

GIVENS, W., *Numerical Computation of the Characteristic Values of a Real Symmetric Matrix* (Oak Ridge National Laboratory, ORNL–1574, 1954).

HOUSEHOLDER, A. S., and BAUER, F. L., 'On Certain Methods for Expanding the Characteristic Polynomial', *Numerische Math.*, 1959, **1**, 29–37.

JACOBI, C. G. J., 'Über ein leichtes Verfahren die in der Theorie der Säcularstörungen vorkommenden Gleichungen numerisch aufzulösen', *J. für die reine und angewandte Mathematik*, 1846, **30**, 51–94.

KRYLOV, A. N., 'O čislennom rešenii uravnenija, kotorym v techničeskih voprasah opredeljajutsja častoty malyh kolebanii material'nyh sistem', *Izv. Akad. Nauk SSSR. Ser.* 7, *Fiz.-mat.*, 1931, **4**, 491–539.

KUBLANOVSKAYA, V. N., 'On some algorithms for the solution of the complete eigenvalue problem', *Zh. vych. mat.*, 1961, **1**, 555–570.

LANCZOS, C., 'An iteration method for the solution of the eigenvalue problem of linear differential and integral operators', *J. Res. Nat. Bur. Stand.*, 1950, **45**, 255–282.

LEVERRIER, U. J. J., 'Sur les variations séculaire des éléments des orbites pour les sept planètes principales', *J. de Math.* (S1), 1840, **5**, 220–254.

NATIONAL PHYSICAL LABORATORY, *Modern Computing Methods* (H.M.S.O., London, 1961).

RUTISHAUSER, H., 'Solution of eigenvalue problems with the **LR**-transformation', *Appl. Math. Ser. nat. Bur. Stand.*, 1958, **49**, 47–81.

TRAUB, J. F., *Iterative Methods for the Solution of Equations* (Prentice-Hall, New Jersey, 1964).

YEFIMOV, N. V., *Quadratic Forms and Matrices* (Academic Press, New York, 1964).

INDEX